Using Simulation Tools to Model Renewable Resources

Janya Chanchaichujit
José F. Saavedra-Rosas

Using Simulation Tools to Model Renewable Resources

The Case of the Thai Rubber Industry

Janya Chanchaichujit
School of Management
Walailak University
Nakhon Si Thammarat
Thailand

José F. Saavedra-Rosas
University of Chile
Santiago
Chile

ISBN 978-3-319-55815-8   ISBN 978-3-319-55816-5 (eBook)
DOI 10.1007/978-3-319-55816-5

Library of Congress Control Number: 2017936921

© The Editor(s) (if applicable) and The Author(s) 2018
This work is subject to copyright. All rights are solely and exclusively licensed by the Publisher, whether the whole or part of the material is concerned, specifically the rights of translation, reprinting, reuse of illustrations, recitation, broadcasting, reproduction on microfilms or in any other physical way, and transmission or information storage and retrieval, electronic adaptation, computer software, or by similar or dissimilar methodology now known or hereafter developed.
The use of general descriptive names, registered names, trademarks, service marks, etc. in this publication does not imply, even in the absence of a specific statement, that such names are exempt from the relevant protective laws and regulations and therefore free for general use.
The publisher, the authors and the editors are safe to assume that the advice and information in this book are believed to be true and accurate at the date of publication. Neither the publisher nor the authors or the editors give a warranty, express or implied, with respect to the material contained herein or for any errors or omissions that may have been made. The publisher remains neutral with regard to jurisdictional claims in published maps and institutional affiliations.

Cover illustration: © Harvey Loake

Printed on acid-free paper

This Palgrave Macmillan imprint is published by Springer Nature
The registered company is Springer International Publishing AG
The registered company address is: Gewerbestrasse 11, 6330 Cham, Switzerland

To my mother, Armui Chanchaichujit, and in memory of my dear father, Anuwat Chanchaichujit, who told me to believe in the power of my dreams and never surrender. Thank you to my wonderful parents. Their faith in me and my capabilities motivates me all the time.
Janya Chanchaichujit

To Paulina and Manuel, my trusted and much loved partners in this wonderful adventure called life...
José F. Saavedra-Rosas

Preface

This book is about the application of simulation tools for the modelling of renewable resources in the Thai rubber industry. It aims to provide a hands-on approach to supply chain modelling by exploring the process and decisions in building a model. The emphasis is on the Thai rubber industry, modelled with the use of an open source software tool, Ptolemy II.

The contributions of this book are twofold with regard to modelling levels: how to model discrete event simulation (DES) using Ptolemy II, and how Thai rubber production, at an industrial level, may benefit from using this model for decision support systems. In addition, the simulation model developed in the book is made accessible to the non-technical reader–the manager, the entrepreneur or the policymaker in the Thai rubber industry—who can replicate it by following the instructions provided and by making use of open source software tools. The benefit of using open source tools is that the reader is not required to make a large investment (apart from their own time) in order to implement and use the models developed in the book. Techniques similar to those described in the book may be used not only for rubber-related problems, but also for any general application in renewable resources management, including the palm oil industry, forestry, farming and livestock management.

This book consists of six chapters. Chapter 1 presents the background related to the book: the natural rubber industry, decision support systems for the rubber industry and a simulation model. In Chap. 2, the elements of the natural rubber industry supply chain are presented. This chapter aims to introduce the reader to the physical transformation of rubber and categories of natural rubber products. The common definitions for each rubber supply chain entity and their relevant processes are presented, along with logistics and marketing activities and decision making. In addition, supply and demand mechanisms are discussed in order to understand how price is formulated in the natural rubber industry. Chapter 3 is concerned with the discrete event simulation (DES) paradigm. The main objective of this chapter is to provide the reader some basic philosophical foundations of the technique, and it also serves as a brief introduction to Ptolemy II, as it illustrates some of the concepts presented by using the software in the context of a simple example. Chapter 4 presents a discrete event simulation (DES) model, which is built step by step using Ptolemy II, and the elements of the model are explained. After the design decisions have been made, the elements are implemented and integrated into the model being built. In each step of the process, checks and control measures are put in place in order to guarantee the validity of the model in a way that resembles unit testing in programming. In Chap. 5, the building blocks developed in Chap. 4 are used to model the Southern Thailand rubber supply chain. After the model is built, it is validated to check the consistency of the observed movement and the material movement at the regional level. The chapter concludes by analysing the capability of the model using case studies. Chapter 6 concludes the book, with summaries and recommendations for future research.

The authors would like to express their sincere gratitude to their respective workplaces for the support provided to them during their work on the book. Major support came from the School of Management at Walailak University, Thailand; the School of Information Systems at Curtin University; the Mining Department at the Faculty of Mathematical and Physical Sciences, University of Chile; and Optika Solutions, Perth, Western Australia. Special appreciation is expressed to

Thai rubber entrepreneur, Mr. Anan Pruksanusak, for his generous support and time in our consultations on the Thai rubber industry.

Finally, the authors wish to express special thanks to their families, who supported them wholeheartedly during the preparation and writing of this book.

Nakhon Si Thammarat, Thailand	Janya Chanchaichujit
Santiago, Chile	José F. Saavedra-Rosas
December 2016	

Contents

1 Introduction 1

2 The Elements of the Natural Rubber Industry Supply Chain 19

3 Discrete Event Simulation Concepts 41

4 A Hands-On Development of a Discrete Event Simulation Model for the Thai Rubber Industry 65

5 Model Implementation and Validation 97

6 Conclusions and Future Research Avenues 133

Appendix A: Farmer Size Distributions 139

Index 155

List of Figures

Fig. 2.1	Categories of natural rubber products	21
Fig. 2.2	Natural rubber supply chain framework	22
Fig. 2.3	Rubber tapping	25
Fig. 2.4	Field latex production process	27
Fig. 2.5	Cup-lump production process	28
Fig. 2.6	Unsmoked sheet production process	29
Fig. 2.7	Primary rubber production process	29
Fig. 2.8	Small trucks for transporting primary rubber products to traders	30
Fig. 2.9	Diagram of the ripped-smoked sheet production process	33
Fig. 2.10	Diagram of the block rubber production process	34
Fig. 2.11	Diagram of the latex concentrate production process	34
Fig. 2.12	Supply-demand mechanism framework of the natural rubber industry	37
Fig. 3.1	System models taxonomy; in grey, the figure presents the box that contains discrete event simulation. Adapted by the authors from (Lawrence and Stephen 2006)	44
Fig. 3.2	Vergil welcome screen	51
Fig. 3.3	Selection of a new graph editor within Vergil	51
Fig. 3.4	An empty graph editor in Vergil	52

List of Figures

Fig. 3.5	Tour of Ptolemy II	53
Fig. 3.6	Queue and server example	53
Fig. 3.7	Queue and server example	54
Fig. 3.8	Setting up the `DE Director`	55
Fig. 3.9	Setting up the `PoissonClock`	56
Fig. 3.10	`PoissonClock` verification	57
Fig. 3.11	Output of the `timedPlotter` actor for model of Fig. 3.10	57
Fig. 3.12	Output of the `timedPlotter` actor for model of Fig. 3.10 after reformatting	57
Fig. 3.13	Format editor for the `timedPlotter`	58
Fig. 3.14	Output of the `timedPlotter` actor for model of Fig. 3.10 with `values` equal to 0,1,2,3	58
Fig. 3.15	`Ramp` actor default parameters	59
Fig. 3.16	Aspect of the model after `Ramp` actor is added	59
Fig. 3.17	Output of the model after `Ramp` actor is added	59
Fig. 3.18	Model after addition of the `Queue` and `Server` actors	60
Fig. 3.19	Output of the two `timedPlotter` actors	60
Fig. 3.20	Model after addition of the `PoissonClock` and `ExponentialDistribution` actors	61
Fig. 3.21	Output of the two `timedPlotter` actors	61
Fig. 4.1	A schematic representation of the supply chain	66
Fig. 4.2	A Vergil graph editor with a DE Director	68
Fig. 4.3	Setting the time in the DE Director actor	69
Fig. 4.4	Setting the time in the DE Director actor	70
Fig. 4.5	Setting the parameters of the clock	70
Fig. 4.6	After dragging the composite actor, it is opened by right-clicking and showing the contextual menu	71
Fig. 4.7	Example of a bounded power law random variable generator in Ptolemy II	72
Fig. 4.8	The interface to define the ports can be accessed by right-clicking the composite actor and selecting `Customise` and then `Ports`	73
Fig. 4.9	Adding a port to a composite actor	73
Fig. 4.10	The template model for each province	74
Fig. 4.11	Iterate Over Array actor implementation logic	76
Fig. 4.12	Output of the execution of the repeat model	77
Fig. 4.13	Iterate Over Array actor implementation logic	78

List of Figures

Fig. 4.14	The SDF Director for the Iterate Over Array actor	78
Fig. 4.15	Example of parameter setting for one `Gaussian` actor	79
Fig. 4.16	Example of a distribution centre	80
Fig. 4.17	Implementation in Ptolemy II of a `Stockpile` composite actor	82
Fig. 4.18	Accessing the Custom Icon menu `Right-Click -> Appearance -> Edit Custom Icon`	83
Fig. 4.19	Custom Icon editor	84
Fig. 4.20	Properties of the white rectangle for stockpile actor animation	85
Fig. 4.21	Inputs to the distribution centre composite actor	86
Fig. 4.22	Implementation of the `Factory` composite actor	88
Fig. 4.23	Implementation of external demand in the model	89
Fig. 4.24	Parameters of Poisson Clock	90
Fig. 4.25	Parameters of Poisson Clock	91
Fig. 4.26	Output of a simple simulation	92
Fig. 4.27	Modification of the actor to add "listeners" for some quantities	92
Fig. 4.28	Output of a simple simulation, outflow from the different stockpiles	93
Fig. 4.29	Output of a simple simulation, evolution of the levels of the stockpiles over time	93
Fig. 5.1	Southern Provinces of Thailand (created by the authors)	99
Fig. 5.2	Nakhon Si Thammarat Province	101
Fig. 5.3	Model used for calibration	102
Fig. 5.4	Detail of the change in productivity to reduce the execution time inside the `IterateOverArray` actor	102
Fig. 5.5	Model implemented for testing total product generation in each Province for 1 year	104
Fig. 5.6	Model implemented including up to distribution centres	106
Fig. 5.7	Modified version of the Factory composite actor that allows distribution of demand between Provinces	109
Fig. 5.8	Modified version of the Factory composite actor that allows distribution of demand between Provinces	110
Fig. 5.9	Implementation of the Dispatcher actor	111
Fig. 5.10	Modified version of the distribution centre composite actor that allows the directing of tokens back to factories	112

List of Figures

Fig. 5.11	An overview of the complete model	114
Fig. 5.12	An overview of the complete model with the extra actors to capture information	115
Fig. 5.13	Graphical output of a run of the model	116
Fig. 5.14	Graphical output of a run of the model with new data	119
Fig. 5.15	Graphical output of a run of the model with new data	122
Fig. 5.16	Graphical output of a run of the model with demand data scaled down by one order of magnitude	123
Fig. 5.17	Changes made to the Dispatcher actor to account for transportation cost from distribution centres to factories	126
Fig. 5.18	Output of the first case study after being run	128
Fig. 5.19	Output of the first case study after being run	131
Fig. A.1	Chumporn farmer size histogram	140
Fig. A.2	Chumporn Province power law fit	140
Fig. A.3	Suratthani farmer size histogram	141
Fig. A.4	Suratthani Province power law fit	141
Fig. A.5	Nakhon Si Thammarat farmer size histogram	142
Fig. A.6	Nakhon Si Thammarat Province power law fit	142
Fig. A.7	Krabi farmer size histogram	143
Fig. A.8	Krabi Province power law fit	143
Fig. A.9	Trang farmer size histogram	144
Fig. A.10	Trang Province power law fit	144
Fig. A.11	Phangnga farmer size histogram	145
Fig. A.12	Phangnga Province power law fit	145
Fig. A.13	Phuket farmer size histogram	146
Fig. A.14	Phuket Province power law fit	146
Fig. A.15	Ranong farmer size histogram	147
Fig. A.16	Ranong Province power law fit	147
Fig. A.17	Songkhla farmer size histogram	148
Fig. A.18	Songkhla Province power law fit	148
Fig. A.19	Satun farmer size histogram	149
Fig. A.20	Satun Province power law fit	149
Fig. A.21	Pathalung farmer size histogram	150
Fig. A.22	Pathalung Province power law fit	150
Fig. A.23	Pattani farmer size histogram	151
Fig. A.24	Pattani Province power law fit	151
Fig. A.25	Yala farmer size histogram	152

Fig. A.26	Yala Province power law fit	152
Fig. A.27	Narathiwat farmer size histogram	153
Fig. A.28	Narathiwat Province power law fit	153

List of Tables

Table 5.1	Rubber production in each Province in year 2015 (adjusted by authors, based on data from (TTRA 2016))	98
Table 5.2	Parameters for modelling farmer size distributions in each Province	100
Table 5.3	Approximate number of farmers in each Province	103
Table 5.4	A comparison of data values and simulated values for farmer production in each Province in 2015	105
Table 5.5	Distribution centre capacity in each Province	107
Table 5.6	Number of factories in each Province	108
Table 5.7	Rubber demand data (adjusted by authors, based on data from (TTRA2016))	116
Table 5.8	Factory demand data (adjusted by authors, based on data from (TTRA2016))	117
Table 5.9	Demand from factories to Provinces (adjusted by authors, data from (TTRA 2016))	118
Table 5.10	New production percentages for each Province	122
Table 5.11	Transportation costs between and within Provinces in baht per kilogram (adjusted by authors, based on data from (Shipping 2012))	125
Table 5.12	Factory demand data for the second case study (adjusted by author, based on data from (DAT 2016))	129

1
Introduction

Abstract This chapter presents the background related to the book: the natural rubber industry, decision support systems for the rubber industry and a simulation model.

Keywords Decision support systems · Simulation model Operations research · Rubber industry

The term "rubber" refers to a highly elastic material which can be stretched to a massive degree without breaking, and which has the ability to return quickly to its original length on release of the stretching force (Barlow et al. 1994). Rubber is a commodity that is used in many products and applications around the world, for both industrial and household use. These range from footwear, to conveyor belts, to sophisticated products such as medical gloves and condoms (Chanchaichujit 2014). The majority of rubber is used in the automobile industry, particularly for tyres. There are two types of rubber: natural and synthetic. Natural rubber is made from the white liquid contents of the rubber tree, whereas synthetic rubber is made in chemical plants using petrochemical products. There are many different types of synthetic rubber products, including emulsion styrene-butadiene rubber (E-SBR), neoprene, nitrile and polyurethane, among many others. However, the most widely used commercial products are generally E-SBR, polybutadiene and neoprene (EIRI 2014).

In the past, natural and synthetic rubber have been considered as product substitutes due to their properties and characteristics. Therefore, changes in the relative price influence the supply and demand of both. For example, when the price of natural rubber rises, the market is more likely to switch to synthetic rubber. Conversely, a rise in petroleum prices may cause a rise in natural rubber demand. However, the interchangeability between synthetic rubber and natural rubber is currently not as strong as it has been in the past, and to assume that this is the case can be risky, as the different modes of production impact the production costs and quality of the various rubber products. Another important factor is customer preference, as the needs of rubber customers can be very sophisticated. Thus, in today's business environment, natural rubber and synthetic rubber are more complementary than substitutive (Barlow et al. 1994). In terms of organisational structure, synthetic rubber production organises itself around large-scale facilities with advanced technology production and is dominated by global enterprises. Since synthetic rubber is a petroleum derived product manufactured in chemical plants, supply and demand management is relatively straightforward. On the other hand, natural rubber is an agricultural commodity which is consumed as an industrial raw material. Over 80% of natural rubber is sourced and processed from millions of tiny independent small farmers with traditional labour. Consequently, rubber becomes a large social commodity when more than 30 million small farms are at stake worldwide (Budiman 2002).

This book will focus on the natural rubber sector due to the importance of this industry to the world economy, and in particular to the Thai rubber industry, which is the world's largest natural rubber producer. Thailand currently produces 4.47 million metric tonnes, nearly one-third of the total global output for 2015 (RRIT 2015). This industry is driving the growth of the Thai economy with revenue generated from rubber and rubber-related product exports valued in 2015 at 12,345 million USD (RRIT 2015). The Thai rubber industry is vital to Thai society, particularly in relation to employment, with approximately 6 million people working in its various sectors (TRA 2007). Social welfare has also greatly improved in Thailand due to the rubber industry's economic contributions. The industry has been instrumental in Thailand's growing economic competitiveness in the prospective world rubber market.

As rubber demand continues to rise, Thailand's rubber plantation areas and manufacturers have grown significantly, resulting in more complexity and uncertainty in decision making. It can be observed that, despite the importance of the natural rubber industry, there are few studies of the industry's supply chain, particularly with regard to the development of a decision support model. In order to gain a competitive advantage and to remain the world leader in rubber production, research is clearly needed to support decision-making capabilities which will allow effective management and enhancement of the supply chain. This book aims to fill the gap by developing a decision support system, particularly with regard to using a simulation tool to assist policy makers and entrepreneurs in their planning and decision making.

1.1 The Natural Rubber Industry

1.1.1 Natural Rubber Demand

Global demand for natural rubber was 12.4 million tons in the year 2015 and it is expected to increase to 14.2 million tons by the year 2020. Demand is expected to grow at an average increase of 3.7% per annum over the next 10 years.

Strong growth in the global automobile industry is expected to drive the worldwide rubber industry with particular demand for rubber from China, India, South Korea and regions in South America (TRA 2010). The tyre industry is the most dominant in terms of rubber consumption, accounting for approximately 70% of the total demand (SRI 2011). In addition, the rubber latex market accounts for 12% of total demand, with its main products being medical gloves. Demand from the latex market is expected to continue over the forecast period, due to the stringency of occupational health and safety regulations and the expansion of the ageing population in the USA, Europe and Japan. Other natural rubber products include shoe soles, with non-tyre automobile components making up the remaining 18% of total demand.

1.1.2 Natural Rubber Supply

Thailand is currently the world's largest natural rubber producer with a worldwide market share of 33%. Indonesia, Vietnam, China and Malaysia rank from second to fifth. Thailand is expected to continue as the leading rubber producer, followed by Indonesia and Vietnam. Malaysia has lagged behind Vietnam and China since 2015.

1.1.3 Price of Natural Rubber

A number of factors affect the price movement of natural rubber, including future market activities, currency movements, weather, and supply and demand factors. However, the fundamental factors influencing rubber prices are supply and demand. The long-term price of rubber depends on technological and economic development, and in the medium-term, rubber price trends will depend on the cyclical effects of the world economy. Short-term factors such as weather, currency exchange rates and rubber trading mechanisms, and government intervention schemes are factors which drive a rise or a fall in rubber prices. However, there are additional fundamental and speculative factors with regard to supply and demand which will have a direct effect on the rubber price at all times (Budiman 2002).

1.1.4 The Thai Rubber Industry

Thailand is currently the world's largest natural rubber producer, producing 4.47 million metric tonnes, nearly one-third of the total global output for the year 2015 (RRIT 2015). Rubber trees were first planted in Thailand in 1899. Rubber production and plantations were later promoted in the southern and eastern regions of the country, and then later further spread into the northeastern region. Since then, rubber plantation areas have multiplied throughout the country. In 2015, rubber plantation areas in Thailand covered a total of 19 million rais, or approximately 3 million hectares (OAE 2015). The vast majority of Thailand's natural rubber is produced by smallholders, who account for almost 90% of rubber production.

1.2 Operations Research Techniques in Supply Chain Management

This book presents the construction and use of discrete event simulation (DES) models for the Thai rubber industry. Despite its specificity, the topic of the book may be extended to the more general topic of the use of tools for supply chain management. Simulation is but one tool in a comprehensive array of tools that belong to the operations research area. For the purposes of giving the reader an idea of the type of techniques that have been used in supply chain management, without going into too much detail, some examples in green supply chain management (GSCM) have been selected. These should enable motivated readers to continue their own research in this exciting area.

1.2.1 Simulation

The simulation technique creates a mathematical model to imitate the behaviour of a real-world process over time. It is used to provide a "what if" analysis of the operation of a system (Terzi and Cavalieri 2004). In supply chain management research, simulation is commonly used for scenario generation and analysis (Srivastava 2007).

An example of research using the simulation method developed for the industrial supply chain network may be found in Khoo et al. (2001). The authors created a simulation model to investigate the economic and environmental impact on the aluminium metal supply chain. The model was used to examine transportation pollution, marketing costs, time to market, recycling of scrap and energy conservation for different choices of location and modes of transportation in the entire supply chain of four production plants. The authors pointed out that to create a reliable and robust simulation system, the model developer has to focus on the accuracy of data, parameters and system behaviour.

Another research group using simulation methods in supply chain management is Teunter and Vlachos (2002). They developed a simulation model to investigate a hybrid production system with manufacturing and remanufacturing under the assumption that "the remanufacturing would be profitable if there is more demand than return products in this system". The simulation model results were then used to analyse costs reduction from different return items. The results in this particular study showed that the disposal process is not necessary unless demand is very low. Adhitya et al. (2011) presented a decision-making support tool to evaluate various environmental impacts on different supply chain network designs. Three scenarios of supply chain decisions that related to changes in inventory composition, distribution network configuration and ordering policy were analysed. The disposable nappy business supply chain was selected to illustrate the methodology. The results obtained from this research found that restructurings in inventory composition benefitted both economic and environmental performance. For network configuration, the results showed that the more distributors added, the higher the environmental impact and costs. In contrast, a less frequent ordering policy was deemed to reduce transportation costs and environmental pollution.

A simulation model developed to analyse the effects of information sharing in the supply chain can be found in Kainuma and Tawara (2006). In their model, a supply chain with customer information and lead-time sharing is examined and compared with a supply chain with no information sharing. The simulation illustrated that a supply chain with information sharing can decrease the bullwhip effect and the out of stock ratio at the retailer.

For the food supply chain viewed from an ecosystem context, Jacxsens et al. (2010) analysed the impact of different logistic system designs on climate change. A simulation model was used to investigate product quality, safety parameters and total costs in each packaging technology in order to anticipate changes in the logistics chain. The climate change impact of this research includes changes in extreme weather conditions, temperature, rainfall, food and waterborne diseases, and the environmental consequences of diminishing the location of crop product areas. The author defined this model as a "complex dynamic ecosystem". Along with the simulation model of the fresh produce logistics chain, an optimisation model to optimise packaging technologies to maintain quality and product safety was presented. The results of this research provided insights towards a scenario analysis of fresh produce biological risk assessment and quality assurance guidelines.

The above research demonstrates how to develop a simulation model in supply chain management research, particularly in green supply chain management. It appears that the simulation method is an efficient tool that allows the decision maker to test the effects of alternative scenarios. However, the simulation model does not generate optimal solutions but evaluates alternative options. Therefore, this method is not applicable for a model seeking "what is best" solutions.[1]

1.2.2 Heuristics

The heuristic method is generally used as a complementary technique for solving mathematical programming models in supply chain management problems. The use of the heuristic method can help to reduce a problem to a manageable size in order to find a better solution (Coyle et al. 2003). However, it does not provide an optimal solution. Mula et al. (2010) reported that heuristics were mainly used in mixed-integer linear programming models and non-linear multi-objective models(MLP) in complex supply chain problems. It is worth noting that the heuristic method has been used more widely for models developed for reverse logistics and waste management (Jayaraman and Ross 2003). Furthermore, Melo et al. (2009) highlighted that the heuristic method appeared to be the only technique suitable for solving the supply chain network problem in dealing with more than one facility layer for reverse logistics.

An example of research utilising heuristic methods for solving supply chain management problems can be found in Jayaraman et al. (2003). They used a heuristic concentration procedure to manage the model's complexity to solve a problem in reverse logistics distribution in order to minimise the total reverse distribution costs. Another research project that adopted the heuristic method for solving a model integrating forward and reverse network flow is that of Ko and Evans (2007). They presented a mixed-integer nonlinear programming model for the design of third-party logistics (3PL) warehousing and transportation operations. In addition, Neto et al. (2008) employed a two-phased heuristic approach to solve their multi-objective linear problem. This model aimed to minimise total costs, cumulative energy demand and waste in a reverse logistics network. The authors mentioned that the heuristic method could be used to overcome the drawbacks of a multi-objective optimisation problem in terms of central processing unit (CPU) time intractability and visual representation for large variables and parameters from a case study.

In the industrial supply chain network modelling approach, the heuristic method has been adopted to solve problems related to facility locations, as evidenced in the work of Lee and Dong (2008). They discussed a logistics network design for computer product recovery to minimise the total costs and the total environmental impact upon the logistics network. Due to the complexity of such a network design problem, a heuristic approach was selected to solve the proposed model, and this incorporated the locations of depots and the construction of a feasible solution for the shipment of products.

1.2.3 Optimisation-Based Techniques

Optimisation-based techniques involve the use of mathematical procedures that are guaranteed to find the optimal solution under a given set of relevant assumptions, constraints and data (Coyle et al. 2003). To accomplish this, the model must incorporate different mathematical programming techniques such as linear programming (LP), mixed-integer programming (MIP), and nonlinear programming (NLP) (Dantzig 2002). Many of these mathematical programming techniques for optimisation models have been incorporated into software packages such as Lingo and CPLEX, which are commercially available for solving large-scale supply chain models. In the existing literature, LP has enabled the modelling of many GSCM problems, as evidenced in the work of Sheu et al. (2005).

The authors presented a linear multi-objective programming model that optimised the operations of an integrated supply chain of computer products. They pointed out that the proposed model conveyed two significant outcomes: (1) a general mathematical model that may be used in any industrial case study and (2) a 21.1% improvement in the operations costs of the supply chain.

A model in the food product industry was created by Soysal et al. (2014), who proposed a linear multi-objective programming model for a generic beef supply chain in Brazil to minimise total logistics costs and greenhouse gas (GHG) emissions. Total logistics costs and GHG emissions are measured by the following activities: inventory, transportation of full and less than fully loaded trucks for road transport, and transportation of other modes such as rail, air and ocean between departure and arrival points. Since LP applicability is limited, given the need for problem formulation to be one of linear approximation, MILP has been used to address this limitation for models dealing with issues related to fixed and variable costs and economies of scale (Coyle et al. 2003). In the MILP formulation, continuous variables are used to represent material flow, while binary variables are used to indicate decisions such as the selection of facilities or the type of transport. This MILP is mainly used to find the optimal configuration for the supply chain network. The use of MIP in the GSCM model can be found in Hugo and Pistikopoulos (2005), Bojarski et al. (2009) and Wang et al. (2011).

1.3 Decision Support Systems for the Rubber Industry

Given the growing complexity and uncertainty in many decision situations, the search for ways of helping managers use quantitative models to support their decision making and planning is an important research topic (Power and Sharda 2007). Decision support model applications have been used in many businesses, ranging from management planning to resource allocation, transportation routing and supply chain management.

Supply chain management problems are generally large-scale and complex in nature. They involve a large number of parameters, decision variables and constraints. Therefore, sophisticated and powerful decision support systems are needed to deal with these problems. Power and Sharda (2007) found that three complex techniques have been widely used to build decision support systems: decision analysis, optimisation and mathematical programming modelling and simulation. Decision analysis involves quantified evaluations of possible alternative options such as decision trees and multiple-criteria decision analysis (Power and Sharda 2007). The optimisation modelling approach is based on a mathematical formulation of the problem to find the optimal solution (Dantzig 2002). The simulation method provides decision makers with tools to evaluate different scenarios (Terzi and Cavalieri 2004). With the use of these techniques, supply chain problems can be addressed through the integration of different activities and criteria, thereby providing managerial insight that can be used to improve the performance of the supply chain. A review of various decision support systems and mathematical modelling techniques used in supply chain problems can be found in Riddalls et al. (2000), Srivastava (2007), Melo et al. (2009), Sasikumar and Kannan (2009) and Chanchaichujit (2014).

Research into the rubber supply chain in Thailand was initiated in 2008 in order to support strategic planning and policy as rubber demand continued to rise. From then on, research studies were conducted by supply chain research working groups such as Wasusri and Chaichompoo (2008), Kritchanchai (2009), Kritchanchai et al. (2010), Chanchaichujit et al. (2016a, b). Research into the literature on the rubber supply chain revealed its working state and contributed to identifying and describing the weaknesses in the chain. Early works tended to focus only on parts of the supply chain and did not adequately investigate the complete supply chain network. For instance, Wasusri and Chaichompoo (2008) studied outbound logistic networks, while Kritchanchai (2009) focused on inbound logistics networks. The complete rubber supply chain, including inbound, manufacturing and outbound distribution, has not yet been studied. In addition to partially detailing the supply chain, a limited number of studies have developed decision support models for policy makers in the Thai rubber industry. A recent work which explored this approach is that of Chanpuypetch and Kritchanchai (2009). They created a decision support model for route selection from Thailand to China. The fuzzy analytic hierarchy process (FAHP) was selected as a tool to create the model. However, the work appeared to develop a model framework rather than developing a decision support model. In Chanchaichujit et al. (2016b), the author developed an optimisation model by formulating costs and GHG emissions as two single objective functions. The objective function of minimising total costs represents economic performance, while minimising total GHG emissions indicates environmental performance. The aim of this work is to provide a decision support tool for policy makers to manage the Thai rubber supply chain. Furthermore, Chanchaichujit et al. (2016a) continued to explore the Thai rubber supply chain by investigating the impact of the restructuring of transportation and distribution on costs and GHG emissions.

Although a number of works have explored the Thai rubber supply chain using decision analysis and optimisation models (Wasusri and Chaichompoo 2008; Kritchanchai 2009; Kritchanchai et al. 2010; Chanchaichujit et al. 2016a, 2016b), there is still no efficient tool that allows the decision maker to test and evaluate the effects of alternative scenarios. From a decision-making perspective, such a tool is crucial when decisions are being made in uncertain and complex circumstances. van der Vorst et al. (2009) notes that simulation tools are always the best choice for supporting decision making in supply chain design and redesign. The research questions in this book have emerged from the gaps in previous research on the Thai rubber supply chain, along with the complex and uncertain concerns of the global rubber industry. The aim is to seek ways to develop decision support tools to enhance Thai rubber industry competitiveness.

1.4 Software Used in the Book

As mentioned in the Preface, Ptolemy II is the tool chosen by the authors to illustrate ways of using DES in modelling Thailand's rubber supply chain. There are several choices in the market for DES packages; Ptolemy II was selected for three main reasons:

- *It is open source*: This allows the reader to follow the modelling without having to spend money on licenses. Ptolemy II uses a very generous Berkeley Standard Distribution (BSD) license that allows anyone to

extend/modify the software and even create commercial derived works without the payment of royalties. The only requisite is the inclusion of the license notice for proper acknowledgement of the use of Ptolemy II in derived works. This book strictly complies with this requirement.
- *The drag-and-drop nature of the graphical user interface*: There are several good open source libraries out there that implement DES models, however, the programming skills needed to undertake such a task are usually well outside the capabilities of researchers and practitioners. As the book's main objective is to introduce the ideas of DES into Thailand's rubber supply chain, it is then desirable to simplify the modelling process. It is the opinion of the authors that Ptolemy II's intuitive GUI design simplifies the model building task.
- *Ptolemy II's maturity level*: The tool was developed by the Center for Hybrid and Embedded Software Systems (CHESS) at the University of California at Berkeley. It is the work of many dedicated scientists over more than 20 years and, in the authors' opinion, incorporates many desirable characteristics that positively differentiate it from other available tools. The tool comes with extensive documentation, and has a large community of users with a dedicated forum where questions regarding specific problems with the use of the software are promptly answered.

Ptolemy II can be downloaded from http://ptolemy.eecs.berkeley.edu/ptolemyII/. There are several versions of the software; we recommend downloading the latest release available (at the time of the writing of this book, this corresponds to version 10.0.1). Documentation for the tool is available in two forms: inside the software and as a freely accessible electronic book (Ptolemaeus 2014), which can be downloaded from http://ptolemy.eecs.berkeley.edu/books/Systems/.

It is highly recommended that the novice user gain some exposure to the documentation and the tool before fully embarking upon a reading of this book. Readers can still follow the book without the need to fully immerse themselves in the tool; however, in order to reap the full benefits of the information contained in these pages, a minimum proficiency with Ptolemy II's GUI and its related concepts is desirable.

1.5 Contributions of the Book

The contributions of this book are twofold with regard to modelling levels: how to model DES by using an open source software tool, Ptolemy II, and how Thai rubber production, at an industrial level, may benefit from using this model for decision support systems. In addition, a simulation model developed in this book is made accessible to non-technical readers, such as managers, entrepreneurs and policy makers in the Thai rubber industry, through easily-understood instructions. The benefit of using open source tools is that the reader can re-apply the model developed in this book to similar decision problems involving other renewable resources, such as the palm oil industry. The contributions are expanded upon below:

- *Modelling-level contribution*: A DES model framework and construction by using the software tool Ptolemy II in a hands-on approach. The model developed in this book also serves as the groundwork for constructing a simulation modelling framework for the reader to apply to similar decision problems. Furthermore, this book uses open source tools, so there will be no licensing barriers for the reader in accessing the software tools.
- *Industrial-level contribution*: Provision of decision support model for non-technical readers such as managers, entrepreneurs and policymakers in the Thai rubber industry, with a view towards increasing the Thai rubber industry's competitiveness.

1.6 Summary

This chapter has provided the general background related to relevant issues in order to clarify and underscore the importance of this book. Based on the existing literature, the chapter addressed the gaps in previous studies of the rubber industry. While the first section of the overview is devoted to background information on the rubber industry, the next section discusses a decision support system for use in the rubber industry, in particular how to apply simulation tools to fill these gaps.

Note

1. In some cases, the development of an effective simulation model may take considerable time and require a high level of programming along with a simulation program package, as mentioned in Khoo et al. (2001).

References

Adhitya, A., Halim, I., & Srinivasan, R. (2011). Decision support for green supply chain operations by integrating dynamic simulation and LCA indicators: Diaper case study. *Environmental Science & Technology, 45*(23), 10178–10185.

Barlow, C., Jayasuriya, S., & Tan, C. S. (1994). *The world of rubber industry*. London: Routledge.

Bojarski, A. D., Laínez, J. M., Epuna, A., & Puigjaner, L. (2009). Incorporating environmental impacts and regulations in a holistic supply chains modeling: An LCA approach. *Computers & Chemical Engineering, 33*(10), 1747–1759.

Budiman, A. (2002). *Recent developments in natural rubber prices* (techreport, FAO).

Chanchaichujit, J. (2014). *Green supply chain model for the thai rubber industry*. Ph.D. thesis, Graduate School of Business, Curtin University, Australia.

Chanchaichujit, J., Saavedra-Rosas, J., & Kaur, A. (2016a). Analysing the impact of restructuring transportation, production and distribution on costs and environment a case from the Thai rubber industry. *International Journal of Logistics Research and Applications*, 1–17. doi:10.1080/13675567.2016.1217317.

Chanchaichujit, J., Saavedra-Rosas, J., Quaddus, M., & West, M. (2016b). The use of an optimisation model to design a green supply chain: A case study of the Thai rubber industry. *The International Journal of Logistics Management, 27*(2), 595–618. doi:10.1108/IJLM-10-2013-0121.

Chanpuypetch, W., & Kritchanchai, D. (2009). Gateway selections for Thailand rubber export. In Asia Pacific Industrial Engineering and Management Society, Kitakyushu, Japan.

Coyle, J., Bardi, E., & Langley, C. (2003). *The management of business logistics: A supply chain perspective*. SWC-Management Series. South-Western/Thomson Learning.

Dantzig, G. (2002). Linear programming. *Operations Research, 50*(1), 42–47.

EIRI. (2014). *Modern rubber chemicals, compounds and rubber goods technology*. Engineers India Research Institute.

Hugo, A., & Pistikopoulos, E. N. (2005). Environmentally conscious long-range planning and design of supply chain networks. *Journal of Cleaner Production*, *13*(15), 1471–1491.

Jacxsens, L., Luning, P. A., van der Vorst, J. G. A. J., Devlieghere, F., Leemans, R., & Uyttendaele, M. (2010). Simulation modelling and risk assessment as tools to identify the impact of climate change on microbiological food safety the case study of fresh produce supply chain. *Food Research International*, *43*(7), 1925–1935.

Jayaraman, V., Patterson, R. A., & Rolland, E. (2003). The design of reverse distribution networks: Models and solution procedures. *European Journal of Operational Research*, *150*(1), 128–149.

Jayaraman, V., & Ross, A. (2003). A simulated annealing methodology to distribution network design and management. *European Journal of Operational Research*, *144*(3), 629–645.

Kainuma, Y., & Tawara, N. (2006). A multiple attribute utility theory approach to lean and green supply chain management. *International Journal of Production Economics*, *101*(1), 99–108.

Khoo, H. H., Bainbridge, I., Speding, T. A., & Taplin, D. M. (2001). Creating a green supply chain. *GMI*, *35*(Autumn), 71–88.

Ko, H. J., & Evans, G. W. (2007). A genetic algorithm-based heuristic for the dynamic integrated forward/reverse logistics network for 3PL. *Computers & Operations Research*, *34*(2), 346–366.

Kritchanchai, D. (2009). Rubber supply chain in north eastern part of Thailand. In Asia Pacific Industrial Engineering and Management Society, Kitakyushu, Japan.

Kritchanchai, D., Somboonwiwat, T., & Chanpuypetch, W. (2010). Supply chain champion in Thailand: A case of rubber to tire industry. In *International conference on logistics and maritime systems (LOGMS 2010)*, Busan, Korea.

Lee, D.-H., & Dong, M. (2008). A heuristic approach to logistics network design for end-of-lease computer products recovery. *Transportation Research Part E: Logistics and Transportation Review*, *44*(3), 455–474.

Melo, M. T., Nickel, S., & da Gama, F. S. (2009). Facility location and supply chain management a review. *European Journal of Operational Research*, *196*(2), 401–412.

Mula, J., Peidro, D., Daz-Madroero, M., & Vicens, E. (2010). Mathematical programming models for supply chain production and transport planning. *European Journal of Operational Research*, *204*(3), 377–390.

Neto, J. Q. F., Bloemhof-Ruwaard, J., van Nunen, J., & van Heck, E. (2008). Designing and evaluating sustainable logistics networks. *International Journal* of *Production Economics, 111*(2), 195 – 208. Special Section on Sustainable Supply Chain.

OAE. (2015). *Natural rubber: Tapping area production: Year 2014–2016.* Office of Agricultural Economics.

Power, D. J., & Sharda, R. (2007). Model-driven decision support systems: Concepts and research directions. *Decision Support Systems, 43*(3), 1044–1061.

Ptolemaeus, C. (2014). *System design, modeling, and simulation using Ptolemy II.* Ptolemy.org.

Riddalls, C. E., Bennett, S., & Tipi, N. S. (2000). Modelling the dynamics of supply chains. *International Journal of Systems Science, 31*(8), 969–976.

RRIT. (2015). *Thailand natural rubber production.* Rubber Research Institute of Thailand.

Sasikumar, P., & Kannan, G. (2009). Issues in reverse supply chain, part III: Classification and simple analysis. *International Journal of Sustainable Engineering, 2*(1), 2–27.

Sheu, J.-B., Chou, Y.-H., & Hu, C.-C. (2005). An integrated logistics operational model for green-supply chain management. *Transportation Research Part E: Logistics and Transportation Review, 41*(4), 287–313.

Soysal, M., Bloemhof-Ruwaard, J. M., & van der Vorst, J. G. A. J. (2014). Modelling food logistics networks with emission considerations: The case of an international beef supply chain. *International Journal of Production Economics, 152,* 57–70.

SRI. (2011). Sri trang agro-industry public company limited prospectus 2011.

Srivastava, S. (2007). Green supply-chain management: A state-of-the-art literature review. *International Journal of Management Reviews, 9*(1), 53–80.

Terzi, S., & Cavalieri, S. (2004). Simulation in the supply chain context: A survey. *Computers in Industry, 53*(1), 3–16.

Teunter, R. H., & Vlachos, D. (2002). On the necessity of a disposal option for returned items that can be remanufactured. *International Journal of Production Economics, 75*(3), 257–266.

TRA. (2007). Support for rubber-smallholders's life quality: Tra president view. Retrieved June 24, 2011, http://www.thainr.com/en/message_detail.php?MID=62.

TRA. (2010). The rubber situation in 2011: Tra president view. Retrieved June 24, 2011, from http://www.thainr.com/en/message_detail.php?MID=136.

van der Vorst, J., Tromp, S., & van der Zee, D. J. (2009). Simulation modelling for food supply chain redesign; integrated decision making on product quality, sustainability and logistics. *International Journal of Production Research, 47*(23), 6611–6631.

Wang, F., Lai, X., & Shi, N. (2011). A multi-objective optimization for green supply chain network design. *Decision Support Systems*, *51*(2), 262–269.

Wasusri, T., & Chaichompoo, A. (2008). A study of logistics system for exporting natural rubber from Thailand to China. In *13th International Symposium on Logistics*. Bangkok, Thailand

2

The Elements of the Natural Rubber Industry Supply Chain

Abstract This chapter the elements of the natural rubber industry supply chain are presented. This chapter aims to introduce the reader to the physical transformation of rubber and categories of natural rubber products. The common definitions for each rubber supply chain entity and their relevant process are presented along with logistics and marketing activities and decision making. Moreover, supply and demand mechanisms are discussed in order to understand how price is formulated in the natural rubber industry.

Keywords Rubber industry · Supply chain management Logistics management · Rubber farmer · Rubber production Rubber price

This chapter gives an overview of the natural rubber industry supply chain. It also outlines the process of the physical transformation of rubber and categories of natural rubber products. The common definition for each rubber supply chain entity and its relevant processes is presented, along with logistics and marketing activities and decisions. The supply and demand mechanisms are also discussed in order to demonstrate how price is formulated in the natural rubber industry.

2.1 The Physical Transformation of Rubber

Natural rubber can be harvested as fresh latex. It is extracted by tapping into a long cut made in the rubber tree, and extracting the white liquid latex contents. This fresh latex can then be processed into primary rubber products that can be divided, according to the method of initial preparation, into three main categories: field latex (LX), unsmoked sheet (US) and cup-lump (CL). Field latex consists of rubber lattices collected from the tapping process. Unsmoked sheet is dry rubber which is made by adding acid to field latex and then rolling it and flattening it into sheet form. Cup-lump is solid dry rubber collected from the tapping cup. These three primary rubber products can be subsequently processed into different intermediate rubber products to eventually produce consumer goods.

Intermediate rubber products include latex concentrate (LCT), ripped-smoked sheet (RSS), air-dried sheet (ADS), high-grade block rubber (H-STR), standard block rubber (STR) and crepe rubber (CRP). (Details of the processing techniques for these products, including types and grades, will be discussed in Sect. 2.5). Latex concentrate is the raw material used for dipped products such as latex examination gloves, surgical gloves, condoms, elastic threads and adhesives. Ripped-smoked sheet and air-dried rubber are used to produce vehicle tyres and industrial rubber parts (Korwuttikulrungsee 2002). Block rubber and high-grade block rubber are the raw materials used for vehicle tyres, rubber parts and high viscosity products such as shoe soles and belts. Different categories of natural rubber products are shown below in Fig. 2.1.

2.2 The Schematic Framework of the Natural Rubber Industry Supply Chain

In general, the rubber supply chain starts at the farm level where rubber farmers produce fresh latex before processing it into the primary rubber products. The products include field latex, unsmoked sheet and cup-lump. They are sold through established local first-tier market traders and then to second-tier traders in sub-provincial regions. Some of the products may go to a third-tier market trader if the farmers are located far from the intermediate rubber processing factories. Primary rubber traders then deliver the primary products down the chain to the factories that process the intermediate rubber products, which include latex concentrate, ripped-smoked sheet, air-dried sheet, high-grade block rubber, standard block rubber and crepe rubber. The products are then delivered to the final rubber processing factories for use as main raw materials in different consumer products (see Fig. 2.1). After the final rubber processing stage, these products will be sold through distribution channels to end users in different markets around the world. Figure 2.2 presents the schematic framework of the natural rubber industry supply chain. In the next section, the common definition for each rubber entity and its relevant process is explained.

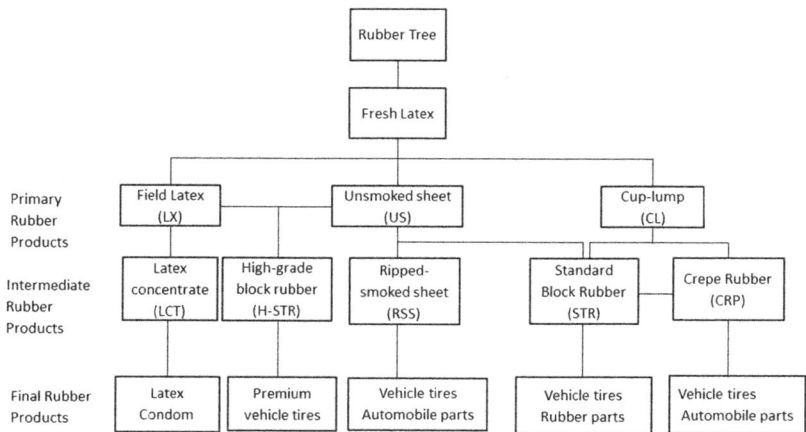

Fig. 2.1 Categories of natural rubber products

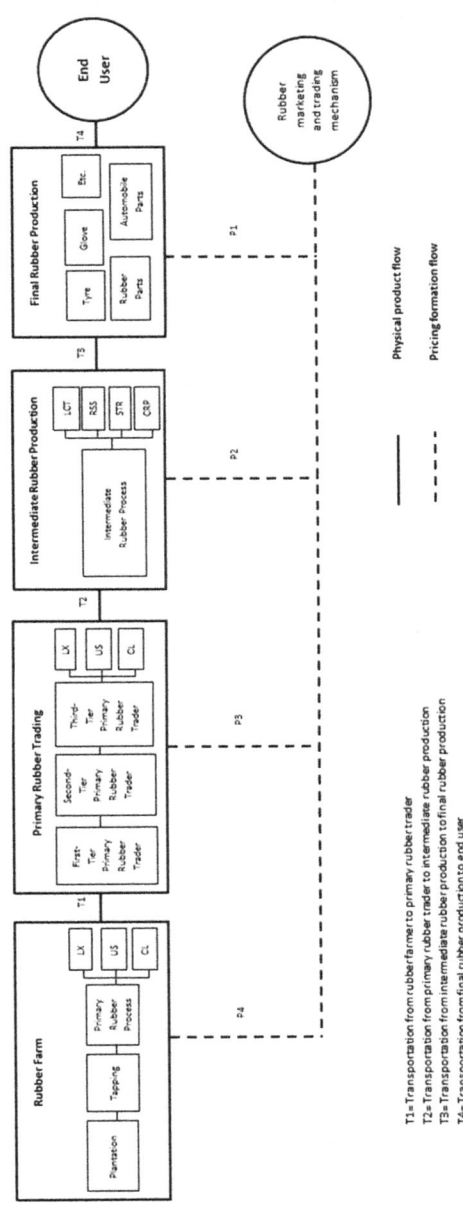

Fig. 2.2 Natural rubber supply chain framework

2.3 Rubber Plantation

2.3.1 Rubber Farmer

Kaiyoorawong and Yangdee (2006), as cited in Chanchaichujit (2014), have defined the rubber farmer as:

- Someone with land rights to plant rubber trees on state land
- Someone who uses their own labour, or that of waged workers, to grow, manage and tap rubber trees
- Someone who is a producer of rubber latex and sheets but who is not involved in high-technology rubber processing or export processing

There are two organisational structures of rubber farm: estate and small holding. Estate rubber farms are run by large investment companies who own land ranging from hundreds to thousands of hectares of planted rubber, and they pay their estate farm labourers in wages. This structure was set up by British investors when rubber plantations in the Malaysian peninsula spread in the early twentieth century (Hagan and Wells 2005). Estate farm producers are served by central processing facilities that are directly linked to rubber goods manufacturers. The structure of estate farms entails activities in vertical integration from upstream to downstream (Barlow et al. 1994). The estate farm therefore has the advantage of bringing in the field latex for further processing and then moving the products downstream for intermediate and final rubber processing for domestic and export markets, with a sizeable economy of scale.

In contrast to the estates, small holding farms are operated by families that produce rubber as an agricultural cash crop. Currently, small holding farmers form the majority in all major rubber producing countries, such as Thailand, Indonesia, Malaysia, Vietnam and China. Thailand, unlike other rubber producing countries, does not have any estate farms. Therefore, Thailand's farm structure comprises only independent smallholders. The smallholding rubber farmer can be classified according to the plantation area the farmer occupies. There are three sizes of smallholding rubber farms; small, medium and large, with areas 25 to 500 rais of land occupation (ORRAF 2012). The proportion of large to small farm size in Thailand is 2%, 8% and 90%, respectively. However, due to the high rubber prices from 2010 to 2012, there has been a dramatic increase in independent farms, which produce rubber as part of their business rather than just a daily cash crop.

Therefore, it can be observed that there has been a shift of smallholding rubber farms from small to medium, medium to large and large to larger of a size up to 5,000 rais. Hence, the current proportions of farm size in some countries such as Thailand and Indonesia have been shifting by 80%, 15%, 10% and 5% of S, M, L and XL, respectively. For smallholdings which use labour to work in rubber plantations, the income is shared with the labourer. The income percentage sharing varies according to each farm's conditions and agreements, which may range from 20% to 50% paid to the labourer. For example, the 80:20 system means 80% of income goes to the owner and 20% to the labourer. Although estate and smallholdings have different organisational structures, they play the same role in contributing to the industry at the farm level to supply primary rubber products.

2.3.2 Planting

The commercial rubber tree *Hevea brasiliensis* contains a latex vessel system in its bark. *Hevea brasiliensis* is a tropical tree, and is best grown in equatorial zones with well-distributed rainfall. After seeding and planting, it takes 5–7 years for a tree to reach maturity before tapping can commence. The planting process starts by clearing uncultivated land in order to prepare the rubber plantation. Budding plants and selected seedlings from nurseries are planted. The density of rubber trees varies from 400 to 700 trees depending on each farm's allocation. As the rubber yield depends on the area planted and the clonal seed of the rubber tree, the planting is a very important stage in ensuring the highest yield to the farmer.

2.3 Rubber Plantation 25

Fig. 2.3 Rubber tapping

2.3.3 Tapping

Tapping is the extraction of rubber liquid from the tree by cutting the bark on the trunk of the tree to sever the latex vessels. It begins when the tree reaches maturity, which takes approximately 5–7 years from initial planting, depending on the conditions of each farm and breeding clone. A cut is made with a knife to allow the latex to flow along the cut and into a cup attached to the trunk (see Fig. 2.3). It takes about 3 h for the latex flow to stop. The extraction of rubber can continue for 15–20 years depending on the standard of tapping. The tapping is normally done on alternate days or on every third day. Some small farmers may tap their trees daily. During winter, when the leaves of the tree die and fall off, its latex yield is reduced and tapping must cease for around 2–3 months. Generally, a rubber tree can be tapped for around 9–10 months a year. The usual procedure in tapping begins with the farmer tapping in the early morning around 3–5 AM, when the rate of latex flow is higher. Each farmer will tap for about 4 hours, which accounts for approximately 300–500 trees depending on the tapping skills and technique. The same farmer will then return to collect the fresh latex from the latex collecting cups on the tree. The latex is then transferred to a bigger container for further processing into primary rubber products.

2.3.4 Primary Rubber Processing

After the fresh latex has been tapped and collected, the farmer has three options for processing the fresh latex into primary rubber products: field latex, unsmoked sheet and cup-lump. These products are sold through the trading system. If farmers decide to sell their product as cup-lump, instead of collecting the fresh latex from the cup in the tapping process, they will add formic acid to the fresh latex in the cup and leave it to dry. The cup-shaped solid rubber, hence the name "cup-lump", is then collected. For field latex, after the fresh latex is collected, the farmer will add ammonia as a preservative before storing the latex in drums or tanks until a buyer is found. For unsmoked sheet rubber, the farmer adds water and formic acid to fresh latex until it coagulates, dries and becomes soft solid rubber.

2.3 Rubber Plantation

Fig. 2.4 Field latex production process

It is then manually flattened with a roller and becomes a white unsmoked rubber sheet. The farmer dries the sheets on a rack in the open air for 1–2 days before selling it along the chain to the intermediate rubber processor for further processing. See Figs. 2.4, 2.5 and 2.6. The whole process is represented in Fig. 2.7.

2.3.5 Logistics

At the farm level, the lead time to transportation of the rubber is relatively short, occurring on a daily basis. Therefore, storage is not an issue at this level. The main modes of transportation for delivering the products to traders are motorcycles and small trucks (Fig. 2.8).

Fig. 2.5 Cup-lump production process

2.3.6 Marketing Decisions

There are three main factors influencing the farmer, farm owner or estate owner in choosing which products they should produce and sell. These include price, labour skill and plantation location. Price appears to be the main factor affecting decisions. However, the price of primary rubber products also depends on the ease of each production. For example, cup-lump requires less activity, less labour and less processing time, which means that the price of cup-lump in general is the lowest in comparison to the other two products. In contrast, the price of unsmoked rubber sheet is the highest due to the method of processing which requires higher labour skills. Therefore, even though price is extremely important, the farmer must trade-off processing time and labour skill against price. Plantation location is another factor affecting processing, in that some plantations are located in remote mountain areas. In these cases cup-lump is the most suitable product as it is easier to transport than liquid latex.

2.3 Rubber Plantation

Fig. 2.6 Unsmoked sheet production process

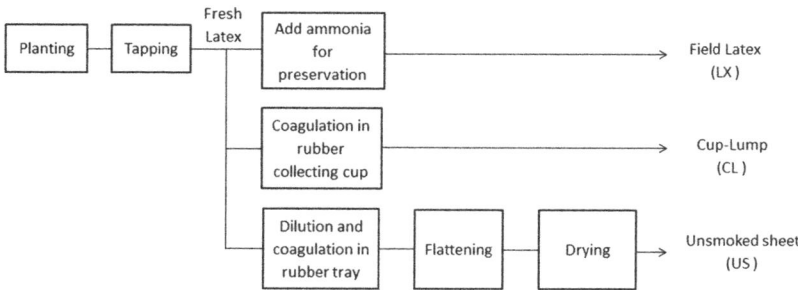

Fig. 2.7 Primary rubber production process

Fig. 2.8 Small trucks for transporting primary rubber products to traders

2.4 Primary Rubber Product Trader

At the primary rubber trading level, trader groups collect and buy primary rubber products from rubber farmers and then sell them down the chain to the manufacturer to process into intermediate rubber products. Primary rubber traders can be classified by network level into first-tier, second-tier and third-tier. The more efficient the network, the fewer trading tiers. In some networks, those that have the capability to implement a vertical integration from upstream to downstream, there may not be a dealer involved. This means that the product is delivered directly from the plantation to the manufacturer. This is the case with estate farms in Malaysia, Indonesia and India, and manufacturing or exporting subsidiary farms in Thailand.

In Thailand, major rubber exporters integrate the upstream business in plantations to secure their supply. This makes them similar to the estate structures that aim to increase the efficiency of the supply chain, reduce costs and increase the quality of products. In contrast to estate and subsidiary farm structures, smallholders with small size plantations run by families which produce rubber as a daily cash crop are heavily reliant on trader networks. The smallholder has to pass up through the tiers of the network until the product is delivered to the manufacturer. Smallholding farmers at the village level can be at a disadvantage in terms of selling their rubber products to dealers. However, governments in all the major producing countries such as Thailand, Malaysia and Indonesia have intervened by establishing cooperatives in villages where there are approximately 30–50 smallholding farmers. In addition, some cooperatives have organised group-processing centres to produce intermediate rubber products. These centres are mainly for ripped-smoked sheet, as the process does not require high-technology machinery. In the process, each cooperative sets up smokehouse facilities to smoke the unsmoked sheet before selling it directly to the exporter. In this way, the cooperative members benefit from producing a value added product (intermediate rubber products) by bypassing the dealer. In Thailand, the government has also established a general market along with cooperatives. This supports the rubber industry by improving trading networks and marketing channels. The general market in Thailand is also called the central rubber market. It serves the industry by being the auction centre for primary rubber products and provides funding, knowledge and techniques related to plantations.

Despite government support, private dealers still dominate in the collection and moving of products down the chain. According to the Thai Rubber Research Institute (RRIT 2015), there were 2553 registered dealers in Thailand. There were 550 rubber cooperatives throughout Thailand in 2011 (ORRAF 2015).

2.4.1 Logistics

First-tier trading is where the lower network traders in each village deal with the collection, handling and delivery of product to either the trader at the next level up or to the manufacturer for further production. As travel distance at this trading level is short, the buying and selling activities are on a daily basis. Farmers deliver their product by motorcycle or four-wheel truck to the dealer facility. In some cases, the dealer may collect the product directly from the plantation. The private dealer will then deliver the product to either their upper level network dealer at a provincial level or directly to the manufacturer. The second and third-tier dealers are larger dealers at the provincial level, and therefore have greater and more efficient means of transport to move the product directly to the manufacturer. At this stage, the traders may keep their private stock instead of selling the product daily. Private stock at this level may be investigated with regard to rubber price speculation as opposed to logistic activities (Chanchaichujit et al. 2016b). In this case, price is driving the buying and selling activities. As mentioned above, the government also plays an important role at this stage in setting up the general rubber market to be the intermediary trader between farmer and factory. The government can intervene by creating buffer stock in the general rubber market in order to manipulate supply and demand and control the price.

2.4.2 Marketing Decisions

In terms of price, lower-tier traders usually offer a lower price to farmers compared to the higher-tiers, due to their proximity to the farmers. This accounts for the lack in economies of scale volume, transportation cost and quality control. The trader's profit comes from the buying and selling price. Therefore it is up to the farmer to judge whether they sell their products to a nearby trader at a lower price or move further up to the higher-tier traders for a better price. A justification has to be made between price and logistics. Likewise for the trader, they can choose whether to sell their products to an upper level trader or directly to the manufacturer.

2.5 Intermediate Rubber Production

Manufacturers process primary rubber products into intermediate rubber products. There are three types of intermediate rubber manufacturing processes:

- Ripped-smoked Sheet (RSS) process
- Block rubber (STR) process
- Latex concentrate (LCT) process

2.5.1 Ripped-Smoked Sheet (RSS) Process

Ripped-smoked sheet is rubber sheet which has undergone the smoking process at a controlled temperature before being classified according to level of quality. There are five grades of RSS products: RSS1, RSS2, RSS3, RSS4 and RSS5. RSS grade is classified by dirt percentage. The higher grades are ranked from RSS1 to RSS5. Higher-grade RSS is mainly used to manufacture pharmaceutical products, while lower grades are used as raw materials in the manufacture of products such as tyres, shoe soles and automobile parts. Figure 2.9 shows the ripped-smoked sheet production activities.

2.5.2 Block Rubber (STR) Process

There are two types of block rubber (STR) products: high-grade block rubber and standard block rubber. Standard block rubber is produced from a mixture of cup-lump, crepe cup-lump and unsmoked rubber sheet. High-grade block rubber is produced from a mixture of field latex and unsmoked rubber sheet. In general, the more field latex in the product, the higher the grade of the block rubber. The amount of production of these two products varies according to customer requirements.

In the block rubber production process, the raw materials are cut and washed before being flattened into a crepe shape. The material then goes through the dryer process to remove excess water and is then compressed into a block shape for storage or shipment. Block rubber is the raw material used in the production of tyres for automobiles and aeroplanes. The block rubber production process is shown in Fig. 2.10.

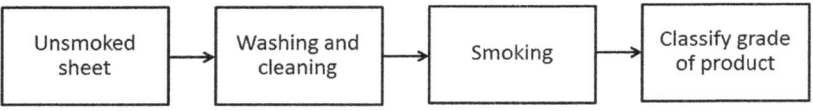

Fig. 2.9 Diagram of the ripped-smoked sheet production process

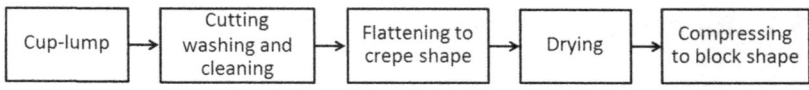

Fig. 2.10 Diagram of the block rubber production process

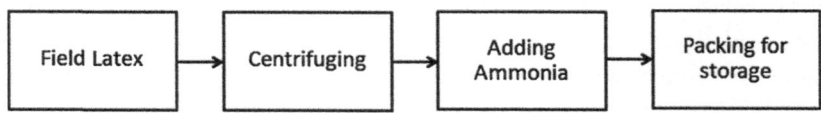

Fig. 2.11 Diagram of the latex concentrate production process

2.5.3 Latex Concentrate (LCT) Process

Latex concentrate is derived from field latex in liquid form. In the latex concentration process, the raw materials are centrifuged to separate out the impurities. Ammonia is then added to prevent coagulation before packing for storage or shipment. Concentrated latex is used primarily in the manufacture of various products such as latex examination gloves, surgical gloves, condoms, elastic threads and adhesives. Figure 2.11 shows the latex concentrate production process.

2.5.4 Logistics

The intermediate rubber market can be regarded as the channel bridging the producers and manufacturers of final rubber products to end-users. It entails the activities of transforming the raw product, further processing, packaging and transport. As mentioned earlier, the transportation from farmer to trader group, and from each trader group to factory, is undertaken mainly by road, as travel distance is short at the provincial level. At the intermediate rubber product stage, transportation from manufacturer to final destination (major rubber consumption markets such as China, India, USA and Europe) comprises different combinations of freight modes such as road, rail and sea (Chanchaichujit et al. 2016a). The logistics system depends heavily on the purchasing contract.

2.5.5 Marketing Decisions

Intermediate rubber marketing decision making is affected by buying, selling and trading; this is how price formation takes place. Decision criteria are affected by trading mechanisms which consist of market types, pricing formula and contract period.

There are three rubber marketing systems: bilateral, spot and exchange. The marketing mechanism is typically activated when the buyer issues a bilateral contract to the seller to lock in long-term sales and purchasing agreements. Once the volume is locked in, sellers can plan and manage their production capacity and inventory balance. The remaining stock volume can be traded on the spot and exchange markets. Under bilateral and spot rubber market systems, the buyer will purchase products from the listed seller with specifications and grade being tested and approved. According to Accenture analysis (Accenture 2014), the natural rubber bilateral trade volume accounted for 75–80% of trade, while spot and exchange trading volume represented 20–25% of total physical rubber trading markets in 2013. In terms of pricing formula, price will be negotiated on a rolling fixed-price basis. However, bilateral long-term contract prices have shifted from a fixed price basis to an indexed price from the exchange market. The indexed price is calculated based on the average closing price over the month prior to delivery. Each seller and buyer may agree to have further premiums or discounts according to quality of the product. The main natural rubber exchanges include the Singapore Commodity Exchange (SICOM), the Tokyo Commodity Exchange (TOCOM), the Shanghai Futures Exchange (SHFE) and the Agriculture Futures Exchange of Thailand (AFET). A natural rubber market structure and price discovery discussion can be found in (Accenture 2014).

2.6 Final Rubber Production

Natural rubber is used to manufacture more than 50,000 types of products (Rubberworld 2012). These range from household goods to transport, construction, recreation and sporting goods, to sophisticated products such as medical gloves and condoms. Products made from rubber include footwear, mats, belts, hydraulic hoses, nets, roofing, toys, lifejackets, thermal insulation and tyres. The German Institute of Rubber Technology (GIRT 2010) reported that world rubber consumption per person, per year, averages 10 to 15 kg. This demonstrates the significance of the rubber industry to the development of the world industrial economy. The rubber goods market can be classified into two groups of products: tyre products and non-tyre products (or general rubber goods). The tyres sector has always been a major rubber consumer. The demand for tyres basically depends on the number of new vehicles, both passenger cars and commercial vehicles, that require new or replacement tyres. Non-tyre products consist of a variety of categories of goods such as manufacturing products (conveyor belts, hoses, tubes, mats, etc.), consumer products (apparel, footwear, sporting goods and equipment), medical goods (gloves, syringes, droppers), and others (seals, rubber-covered fabrics). The development of natural rubber goods production and the tyre manufacturing process can be found in Barlow (1997). Tyre production and analysis has also been broadly discussed in Smith and Burger (1992).

2.7 Supply-Demand Mechanisms of the Rubber Industry

Figure 2.12 illustrates the supply-demand mechanism framework of the natural rubber industry. It can be seen that supply and demand drives the formulation of the rubber price in selling and buying activities.

2.7.1 Supply Factors

There are five major supply factors. These include the rubber price, plantation area, factory processing capacity, input price to produce rubber and technology. In the short term, the price of rubber is a key factor affecting the rubber supply through its profitability, while the price of other types of rubber is also a key factor indicating the attractiveness of choosing what product to produce. However, it must be noted that switching types of rubber production is not only subject to price but also to farmer

2.7 Supply-Demand Mechanisms of the Rubber Industry

preference, which is driven by the ease of processing and farmer skill. As natural rubber plantations are still very labour-intensive, the ease of processing and labour skills can limit the ability to change the production type to other types of natural rubber, even though the price is different. Apart from farmer preference, weather and government intervention are two other elements that influence the price of natural rubber. Government intervention programs, such as the replanting program, subsidise rubber replanting for the small holding farmer and establish group processing for farmers to produce intermediate rubber products. The most important long-term factors are the supply, the price of natural rubber and the input price of each type of natural rubber. The costs of land, capital and labour are direct costs which affect long-term plantation and processing capacity investment. The plantation area can be used to determine the level of the primary rubber supply before differentiation into varying types of intermediate rubber through the processing stage. However, the specific level of rubber supply is established through the processing capacity for different types of intermediate rubber. This is the case for products such as latex concentrate, smoked rubber sheet and block rubber. Technology is another factor influencing supply through productivity, quality and cost.

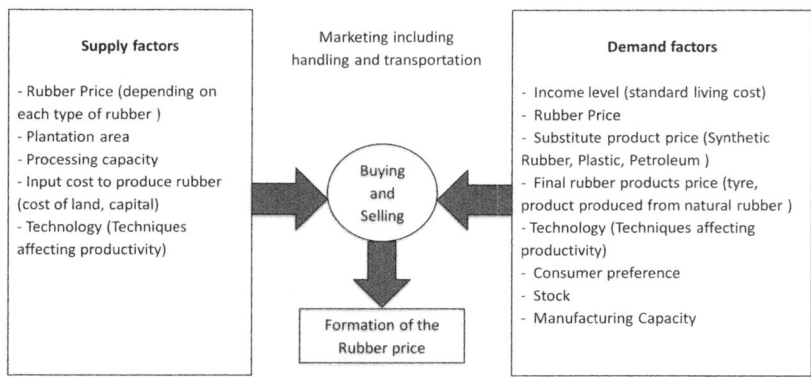

Fig. 2.12 Supply-demand mechanism framework of the natural rubber industry

2.7.2 Demand Factors

Before discussing demand factors in natural rubber, it is important to mention again that natural rubber products are intermediate goods used in producing final consumer goods from gloves to tyres. Therefore the demand for natural rubber depends on many factors influencing not only intermediate goods but also the demand for final goods. An example of this is that tyres as a final product depend very much on the demand for cars, which in turn depends on economics and the income of the consumer. As such, it can be seen that the factors influencing demand are more complex than that of supply. However, Barlow et al. (1994), have classified the main factors influencing demand as follows: income level, rubber price, substitute production price, final rubber product price, technology, consumer preference, stock and manufacturing capacity. With these influences, the level of stock and manufacturing capacity can be seen as short-term factors, as they can be used to limit the manufacturing output responses to increased demand. General income and the price of final goods are categorised as long-term factors. General income can be considered the major factor, as it reflects the world economic situation. General income impacts on the price of the final goods and on the price of rubber and its substitution.

2.7.3 The Formulation of the Rubber Price

Prices are determined by the interaction of supply and demand. As illustrated in Fig. 2.12, the supply of rubber is determined by its price in the marketplace, together with production capacity, input costs and technology. Rubber supply interacts and has a direct relationship with demand, which is influenced by price and income level in the overall economy, the price of rubber substitution, the price of final goods, technology, consumer preference, stock and manufacturing capacity utilisation. The organisational structure of production, marketing and consumption, and government measures around rubber also influence supply and demand but work through the formulation of price.

2.8 Summary

This chapter has given an overview of the natural rubber industry and outlined the process of the physical transformation of natural rubber products. The natural rubber supply chain framework presented in the chapter illustrates the complexities of this industry. It can be seen that the rubber industry starts at the farm level with millions of rubber farmers, and continues through a variety of trading channels into the manufacture of thousands of rubber products. The products are then processed into final consumer products in different regions around the world. In addition to complex physical transformation, logistics and marketing decisions, this industry is also influenced by many factors such as economics, capacity, stock level and technologies. Due to this complexity, a decision support tool that aids efficient decision making in this industry is clearly needed.

In the next chapter, the discrete event simulation (DES) technique will be introduced. The chapter provides the background on how to use this tool to model the rubber industry supply chain before actual model building commences.

References

Accenture (2014). *Extracting value from natural rubber trading markets: Optimizing marketing, procurement and hedging for producers and consumers.*

Barlow, C. (1997). Growth, structural change and plantation tree crops: The case of rubber. *World Development, 25*(10), 1589–1607. doi:http://dx.doi.org/10.1016/S0305-750X(97)00059-4.

Barlow, C., Jayasuriya, S., & Tan, C. S. (1994). *The world of rubber industry.* London: Routledge.

Chanchaichujit, J. (2014). *Green Supply Chain Model for the Thai Rubber Industry.* phdthesis, Graduate School of Business, Curtin University, Australia.

Chanchaichujit, J., Saavedra-Rosas, J., & Kaur, A. (2016a). Analysing the impact of restructuring transportation, production and distribution on costs and environment a case from the thai rubber industry. *International Journal of Logistics Research and Applications*, pp. 1–17. doi:10.1080/13675567.2016.1217317.

Chanchaichujit, J., Saavedra-Rosas, J., Quaddus, M., & West, M. (2016b). The use of an optimisation model to design a green supply chain: A case study of

the thai rubber industry. *The International Journal of Logistics Management,* 27(2), 595–618. doi:10.1108/IJLM-10-2013-0121.
GIRT (2010). Rubber research 2010. German Institute of Rubber Technology. http://www.dikautschuk.de/english/forsch/index.html.
Hagan, J., & Wells, A. (2005). The British and rubber in Malaya. University of Woolongong Research Online (Faculty of Arts-Paper (Archive)).
Kaiyoorawong, S., & Yangdee, B. (2006). Rights of rubber farmers in thailand under free trade: Project for ecological awareness building. Bangkok.
Korwuttikulrungsee, S. (2002). *Natural rubber production.* Pattani Campus: Prince of Songkhla University.
ORRAF (2012). Farmer development. office of the rubber replanting aid fund. http://www.rubber.co.th/ewtadmin/ewt/rubber_eng/ewt_w3c/ewt_news.php?filename=index&nid=1209.
ORRAF (2015). Farmer development. office of the rubber replanting aid fund.
RRIT (2015). Rubber statistics. Rubber research institute of Thailand.
Smith, H., & Burger, K. (1992). *The outlook for natural rubber production and consumption.* Amsterdam: Elsevier.

3

Discrete Event Simulation Concepts

Abstract This chapter is concerned with the discrete event simulation (DES) paradigm. The main objective of this chapter is to provide the reader some basic philosophical foundations of the technique and it also serves as a brief introduction to Ptolemy II as it illustrates some of the concepts presented by using the software in the context of a simple example.

Keywords Discrete Event simulation (DES) · Stochastic modelling Ptolemy II

This chapter presents the concepts used in discrete event simulation (DES) models. The main objective here is to provide a basic introduction to the area and to provide the interested reader with pointers as to where to obtain more information.

3.1 The Nature of Simulation

Simulation can be defined as

> "something that is made to look, feel, or behave like something else especially so that it can be studied or used to train people". (Merriam-Webster 2016).

Another related definition of simulation provided by the same dictionary is

> "the imitative representation of the functioning of one system or process by means of the functioning of another <a computer *simulation* of an industrial process>"

It is clear from these dictionary definitions that simulation is the imitation of real-world processes or systems over time. In order to accomplish a simulation, there is a need to create an artificial history of the system being simulated. Based on that history, conclusions can be drawn about the behaviour of the real-world process/system and inferences can be formulated.

Among many uses, simulation can be used as an analysis tool for existing systems or as an aid in the design and forecasting of the performance of new systems.

In the ideal world, systems should be described by means of a set of mathematical expressions, and the resulting equations used to infer the behaviour of the system. However, some systems, due to their complexity, are not well suited to a mathematical description, for example where the use of numerical models is important to allow a numerical representation of a real-world system, which in turn can be used to perform controlled experiments to analyse system behaviour.

Simulation has a variety of uses. However, there are instances where simulation either is not necessary or should not be used; these are detailed below (Jerry et al. 2010):

- If the problem can be solved by common sense
- If the problem can be solved analytically
- If it is easier to perform direct experiments
- If the costs exceed the savings
- If the resources or time to perform simulation studies are not available
- If no data, not even estimates, are available
- If there is not enough time or personnel to verify/validate the model
- If managers have unreasonable expectations, such as overestimating the power of simulation
- If system behaviour is too complex or cannot be defined

One big advantage of simulation is that it can be used to mimic a system. It can be used to forecast system behaviour without the need for actually building the system or for engaging in expensive experiments. In addition, as representations of real-world systems, simulation models are usually free from mathematical assumptions that are required and the mathematical models can be solved.[1] Of particular interest is the use of simulation models to test "what if" questions regarding systems or processes.

There are some disadvantages related to the use of simulation models. The building of simulation models requires specialist knowledge and it can be a time-consuming task. Furthermore, building a model does not guarantee that the most appropriate decision will be taken. It is up to the analyst to decide which information produced by the model will be used and how that information will be used in the decision-making process.

A system can be *deterministic* or *stochastic*. A deterministic system is one which does not have random components. This means that the state of a system at any given time is known along with the equations that describe its evolution, so it is possible to describe any future state of the system. This view of systems is prevalent in classical mechanics where the models do not have "noise" and whose behaviours are predicted from initial conditions and constitutive equations. On the other hand, a system is stochastic if the randomness present in some of the variables that describe the system cannot be neglected and there is a substantial influence on the future behaviour of the system, causing predictions based in constitutive equations to be unreliable as the system exhibits erratic behaviour.

A system model can be *static* or *dynamic*. A static model is one in which the time variable does not play an important role. A typical example of a static model is the equilibrium model used in economics, where economic behaviour is summarised in the supply and demand curves. In this model the equilibrium is at the intersection of the two curves, which are assumed not to change under the time horizon of the study. If there are changes to any one of the curves (or both of them) then a new equilibrium is reached and will remain this way until a new change is studied. A dynamic model is one in which the time variable plays an important role. For example, if we are interested in the probability that a roller belonging to a conveyor belt system fails, then this probability will depend on how much time has elapsed since the last time the roller was replaced or maintained.

Finally, a dynamic system model, can be *continuous* or *discrete*. A continuous system is described by variables that continuously evolve over time such as position, speed, etc. A discrete system is described by variables that change at specific instants in time and are considered piecewise constant functions. For example, the number of customers in a queue is a discrete variable because it changes at the time when customers enter or leave the queue but remains constant in between. Figure 3.1 summarises this taxonomy of system models.

From a taxonomic point of view, a *discrete event system* is characterised as being stochastic, dynamic and discrete. The model just needs to have at least one system variable that has randomness or is random, the time evolution of the state-variables is important, and the changes of the state variables happen at discrete points in time.

A model for a system requires a proper delineation of what a system is. It can be said that a system is a group of objects that interact or among which there is some interdependence. A simple example of a system could be an open pit mine, which can be defined as simply a collection of mining equipment (together with the deposit itself), which defines the mining system. Whatever is not part of the system but somehow could have an influence on or interact with the system is termed the *environment*. A clear boundary between the system and its environment is important; in some

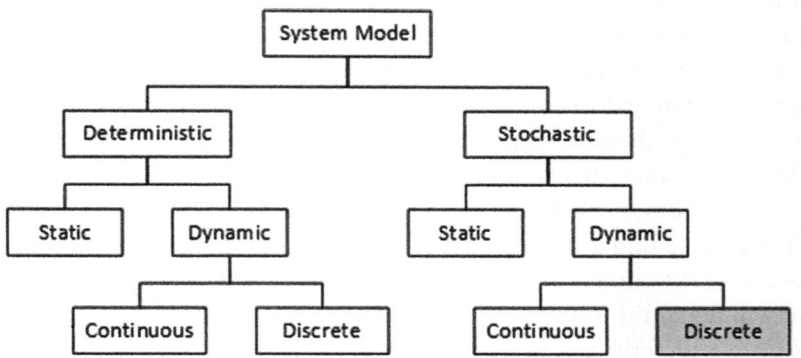

Fig. 3.1 System models taxonomy; in grey, the figure presents the box that contains discrete event simulation. Adapted by the authors from (Lawrence and Stephen 2006)

cases, boundary definition is not a trivial task at all. To illustrate the point, consider the previous example of the "mining" system: Are the miners part of the system? Are the miners' families part of the system? There could be reasons to consider the miners and their families as belonging to the system or reasons not to consider them. In any case, it can be clearly seen that the definition of the system will affect the ways in which the system is described.

An *entity* is an object of interest within the system. In the "mining" system example, one possible entity will be a particular truck belonging to the mine fleet. An *attribute* is a property of an entity. Following the same example, number of tyres is an attribute of the truck entity. An *activity* represents a time period of specified length, for example for the "mining" system an activity can be loading a truck with mineral. The *state* of a system is defined as a collection of variables that can be used to describe the system at any given time in relation to the entities of the system or the objectives of the study. For example, in the "mining" system example, the number of tonnes extracted is one variable that helps to describe the system. Another could be the number of trucks waiting in the loader queue, etc. Finally, an *event* is an instantaneous occurrence that changes the state of the system. For example, for the already familiar "mining" system example, the loading of a truck is an event that changes the state of the system, as once loaded, the truck leaves the loader, giving the opportunity for another truck to start loading, thus changing the length of the loader queue.

A typical simulation study consists of the following steps (Jerry et al. 2010):

1. Problem formulation
2. Setting objectives and overall plan
3. Model conceptualisation (happens in parallel with 4)
4. Data collection (happens in parallel with 3)
5. Model translation
6. Verification (if the model is not verified, go back to 5)
7. Validation (if the model is not validated, go back to 3 and 4)
8. Experimental design
9. Production runs and analysis
10. Additional runs (if more runs are needed, go back to 8 and 9)

11. Documentation and reporting
12. Implementation

In the preceding list, the first step cannot be overstated. A bad formulation of the problem will end up producing a model that could be correct but not comprehensive enough to solve the problem that needs to be solved. The second step is equally important, as it defines the expectations associated with the model's output. The model conceptualisation and data collection steps usually happen in parallel, as there is no point in conceptualising a model for which there is no data and there is no point in collecting data that is not needed by the model. Model translation is the implementation phase of the model into a programming language or software system. It requires verification to check that the output of the model is correct (i.e. producing what it is expected to produce), and it must be validated, meaning that the model is conversant with reality (i.e. that the output of the model checks against data observed in the real world for a certain set of input values). If the model is properly validated, then it is possible to proceed to the experiments which need first to be designed, after which the runs are executed. Finally the whole process is documented and eventually reported to the interested parties. The last step is usually up to the commissioning agency/party to perform. It essentially implies the use of the conclusions obtained in the model to produce a difference in the real world. This is done by means of modifying current operations in light of the simulation outcomes or by proposing a design based on the simulation results.

Random numbers are an essential ingredient in the simulation of discrete systems. Most computer languages or applications have a subroutine, object or function that generates at least *uniformly distributed* random numbers; from there, other distributions can be obtained by using different techniques.

These "generators" create a sequence of numbers $R_1, R_2, \ldots$ that should be *independent* of each other and whose histogram matches that of a uniform histogram. Unfortunately, the algorithms used to generate uniform random numbers generate numbers that are not really independent from one another, and for this reason those numbers are called *pseudo*-random numbers. One very common generator is what is termed *linear congruential*, which essentially produces a sequence of integers $X_1, X_2, \ldots$ between zero and $m - 1$ by using the following recursive relationship:

$$X_{i+1} = (aX_i + c) \mod m \quad \forall i \in \mathbb{Z} \tag{3.1}$$

In Eq. 3.1, the value X_0 is called the *seed*, a is called the *multiplier*, c is called the *increment* and m is called the *modulus*. The selection of the values X_0, a, c, m have a serious effect on the way the generator operates. For example, if it happens that during the generation of the sequence, the value X_0 is reached, then the whole sequence will start to repeat itself; the number of steps before this happens is called the *cycle* length.

Once the sequence of numbers $\{X_i\}_i$ has been generated, it needs to be transformed from an integer to a number in the interval $[0, 1]$. It should be noted at this point in the discussion that the mod operation produces values in the set $\{0, 1, \ldots, m - 1\}$ only. From this point of view, then, it is easy to observe that the quantity

$$R_i = \frac{X_i}{m} \quad \forall i \in \mathbb{Z}$$

produces a set of numbers in the desired interval. Also, it is not difficult to observe that the number of points R_i that can be generated is finite and equal to m. Then, ideally, the value m should be large enough to produce many numbers in the interval $[0, 1]$ (typical values used are $m = 2^{31} - 1$ or $m = 2^{48}$).

What is important to note is that random number generators (RNGs) create a sequence that starts with a seed as an input to the RNG. There are several simulation books that cover the topic of random number generation in more detail (Jerry et al. 2010; Lawrence and Stephen 2006; Law 2015).

3.2 Discrete Event Simulation (DES)

As was defined previously, a discrete event system is characterised as being stochastic, dynamic and discrete. Unfortunately, this definition does not really explain how DES models are implemented. The objective of this section is to provide some insight into the underlying computation model for implementing DES system models.

Heidergott et al. (2010) defines a DES model as a model that generates a sequence $\vec{x}(k)$ of states, starting with an initial state $\vec{x}(0) = \vec{x}_0$. Under the DES paradigm, the state $\vec{x}(k + 1)$ is obtained from $\vec{x}(k)$ by a state-transition mapping τ (measurable) that takes a vector of independent random variables $\vec{y}(k) = (y_1(k), .., y_m(k))$ and

$$X(k + 1) = \tau(\vec{x}(k), \vec{y}(k)) \quad \forall k \geq 0. \tag{3.2}$$

Based on this definition, it is clear that the dynamic aspect of the system is modelled trough the sequence of states $\vec{x}(k)$ (here k plays the role of time), the stochastic aspect of the model is given by the vector $\vec{y}(k)$ and the discrete event is given by the transitions that happen at discrete points in time $k \geq 0$.

The solution to the problem of how to model the transitions described by Eq. 3.2 is to adopt a simulation paradigm that allows the programmer to understand the computer objects involved in running the simulation. Several world views have been developed for DES programming, the most popular ones being Jerry et al. (2010):

- Activity-scanning
- Event-scheduling
- Process-interaction

These different world-views will be briefly described in the next subsections. The reader should be aware that every world view has it pros and cons and that there is no clear way to say which one should be used. Most likely, by choosing a particular software package, the user will implicitly adhere to one world view. Thus, it is important that the reader gains an idea of the most common world views and the types of impacts they can have in the execution of the simulation.

3.2.1 Activity-Scanning World View

This world view focuses on activities and the conditions that allow an activity to begin. In this world view, time is broken down into small increments; at each time point (or clock advance) the model checks the conditions for each activity, and those that satisfy the conditions are then started. It can be noted that this approach could become extremely slow if the time increment is too small. Unfortunately, large time increments will not provide a proper representation of how the activities should be executed, as some activities that satisfy the conditions will need to wait for the next clock advance in order to start. It also needs to be noted that if the time increments are too small, then a lot of the activity checks will be useless as there will be no change in the system at all. For example, if the time is subdivided into 0.01 s increments and there is a task that could start on $t = 10$ then this implies 1000 checks of commencement conditions for all the activities that need to be considered. This may not be that problematic, given today's computing power, but it can be seen that this could easily get out of control if there were too many activities, or if the time increments were much smaller. This type of approach could lead to simulations that could take several hours or even days to run.

3.2.2 Event-Scheduling World View

This world view focuses on events and their impact on the system state. The time advance process is performed based on an ordered list called the *future event list* (FEL). When a new event is generated into the system, it is inserted in chronological order into the list and then the clock moves to the next time corresponding to the first element of the FEL, processes the event, modifies the state of the system and introduces all the changes needed into the FEL. It needs to be noted that the insertion operation could become expensive if there are too many future events that are generated and need to be inserted into the list. Nevertheless, this world view is definitively faster than the activity-scanning view, as it avoids the unnecessary checking of activities at every clock advance.

3.2.3 Process-Interaction World View

This world view focuses on processes instead of activities. A process is a collection of time-sequenced events, delays and activities that define the life cycle of an entity as it moves through the system. With this world view it is possible to have many processes active at any given time and the interaction between processes could become quite complex. This approach produces a more modular code and is considered to be better from a programming point of view.

3.3 Ptolemy II

As mentioned in Chap. 1, this book uses Ptolemy II, one of many available tools developed at the Center for Hybrid and Embedded Software Systems (CHESS) at the University of California at Berkeley. The group is led by Edward E. Lee. Ptolemy is defined as follows (Ptolemaeus 2014):

> "Ptolemy II is an open-source simulation and modeling tool intended for experimenting with system design techniques, particularly those that involve combinations of different types of models. It was developed by researchers at UC Berkeley, and over the last two decades it has evolved into a complex and sophisticated tool used by researchers around the world."

In order to start Ptolemy II,[2] we need to open Vergil, the graphical editor for Ptolemy II, so that the window shown in Fig. 3.2 will appear.

From the initial Vergil window the user needs to create a new Graph Editor, following the sequence `File -> New -> Graph Editor`, this sequence is illustrated in Fig. 3.3.

Finally, a new fresh graph editor is open as shown in Fig. 3.4. This empty editor is the canvas that we use to "draw" our model, which is explained as we progress.

3.3 Ptolemy II   51

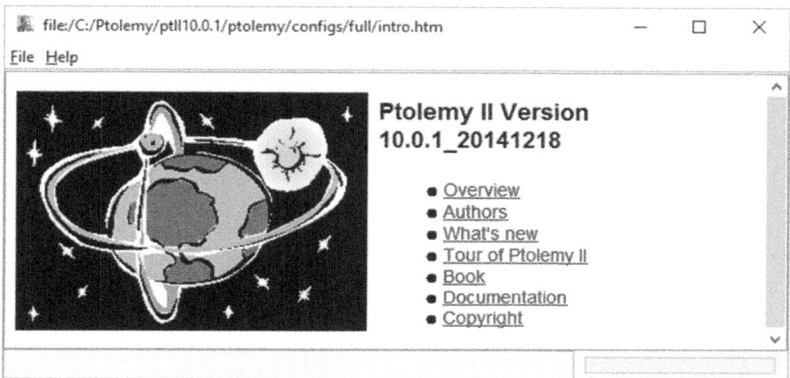

Fig. 3.2 Vergil welcome screen

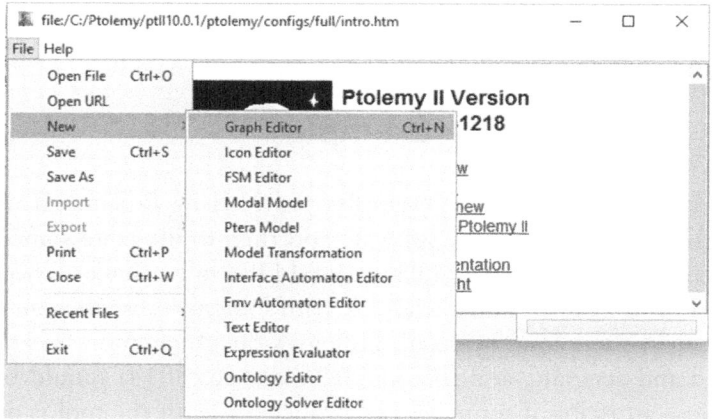

Fig. 3.3 Selection of a new graph editor within Vergil

In order to gain a better understanding of the software capabilities, the next section looks at an example of a model already built in Ptolemy II. For a more detailed description of all the different domains, actors and capabilities of Ptolemy II, please refer to the book (available online) provided by the Ptolemy IIs team (Ptolemaeus 2014).

Fig. 3.4 An empty graph editor in Vergil

3.4 An Illustrative Example

This section uses one of the examples provided by Ptolemy II, first to introduce the features of the software and then to illustrate some of the previous discussion about simulation, and DES in particular. The idea is to select one of the examples available with Ptolemy's distribution and to discuss it in some level of detail. Ptolemy's documentation is extremely detailed and accurate, and the book (Ptolemaeus 2014) should be read with close attention if some level of proficiency with the tool is desired. Once Vergil is launched, the examples that come with Ptolemy can be accessed by clicking on `Tour of Ptolemy II`, which gives access to the window shown in Fig. 3.5.

From here the examples can be selected, and it is recommended that the reader have a look at the different domains and capabilities illustrated by the extensive list provided. The first example considered is in the *Discrete Event Modeling* section under the *Basic Modeling Capabilities* heading of the list of examples. The model is named `QueueAndServer` and is presented in Fig. 3.6.

3.4 An Illustrative Example

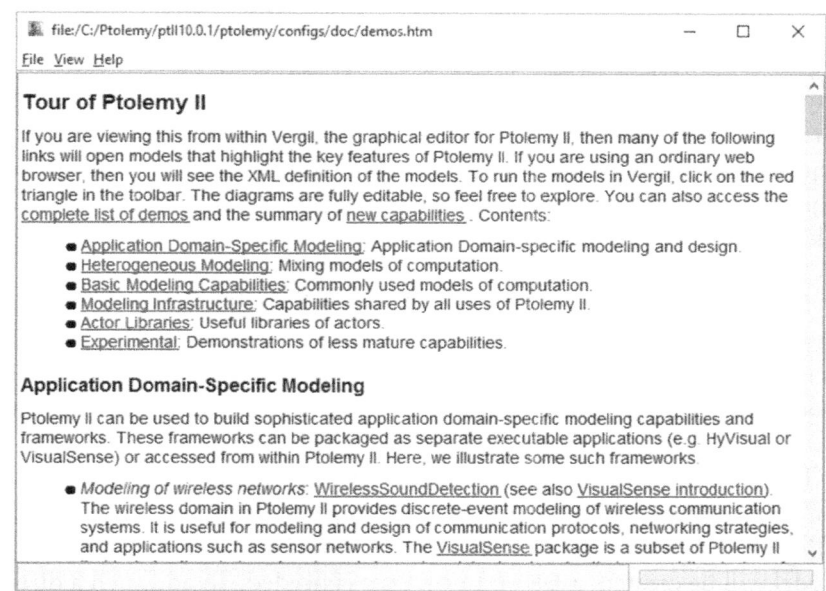

Fig. 3.5 Tour of Ptolemy II

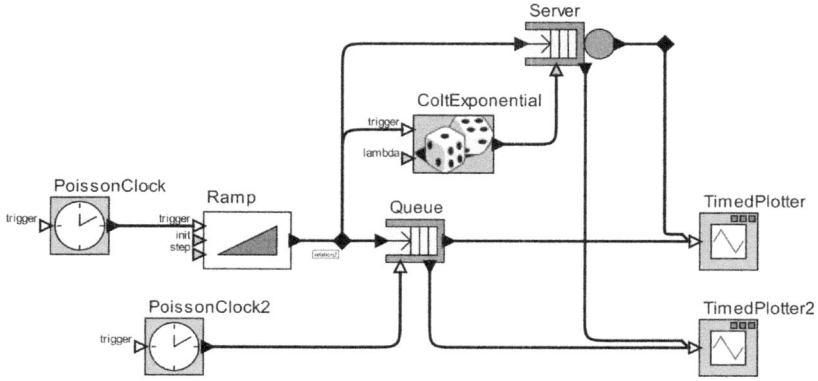

Fig. 3.6 Queue and server example

The idea of this model is to illustrate two different ways of creating an $M/M/1$ queue. An $M/M/1$ queue is characterised as being composed of a single server and the arrival process being a Poisson one, with the service process following exponential service times. Ptolemy has two different actors that have the capability of queuing tokens: Server and Queue. The main difference between the two actors is that the Server actor requires a service time to be passed to it, while the Queue actor instead requires a token to be passed to a port to signal the release of the oldest token present in the queue. If no tokens are in the queue, then the actor fires nothing.

The model is essentially divided into two branches: one branch that uses a Server actor and another that uses a Queue actor. In order to understand the model, we will implement it following a step-by-step process that will help us understand the model-building process in Ptolemy II.

The first thing that needs to be done with the model is to add a "director". The director is in charge of "running" the simulation. There are different directors, each associated with a different domain. As a DES model will be created, a DE Director is needed. To add such a director, it is simply dragged and dropped it into the now empty "canvas", as shown in Fig. 3.7.

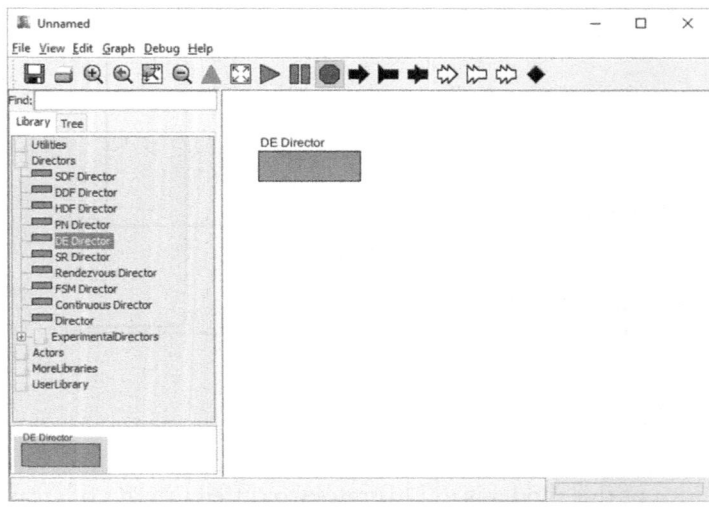

Fig. 3.7 Queue and server example

3.4 An Illustrative Example

The director needs to be set up to proceed with the simulation runs. In particular the stopping time of the clock needs to be specified. The properties of the actor are accessed by double-clicking on it, which in turn presents the dialog shown in Fig. 3.8. Any number can be put into the stopTime field. For illustration purposes a value of 100 will be used.

The next step is to create tokens in the system. This can be accomplished by using source actors. There are GenericSources, Timed Sources and SequenceSources. A more detailed description of the different types of actors can be found in the Ptolemy II book (Ptolemaeus 2014). For the purposes of explaining the model we will proceed to the particular actor chosen for the example. The actor used to generate tokens is a PoissonClock, which can be found in Actors -> Sources -> TimedSources -> PoissonClock. By using a drag-and-drop process, the actor can be added to the model; by double-clicking on it, its properties can be accessed as shown in Fig. 3.9.

It can be observed that the PossionClock parameters are already populated with some defaults. The most important parameters with which to play are seed, meanTime and values. The seed parameter allows the modeller to specify the seed to be used with the RNG. The meanTime parameter is used to specify the average time between events in a Poisson distribution. A Poisson distribution models a random variable for which the time between events follows an exponential distribution with mean value λ. The values parameter specifies the list of values to be used

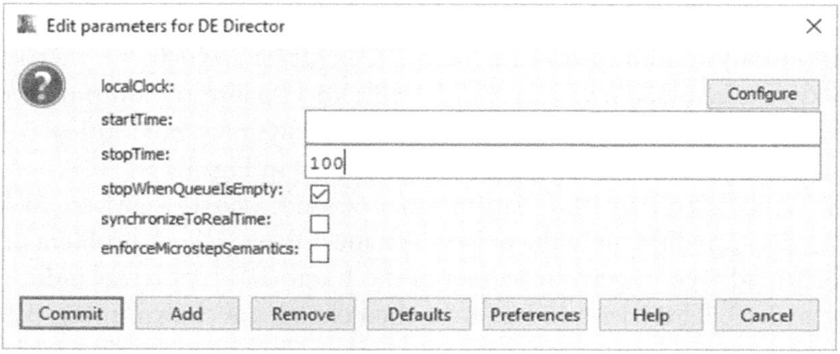

Fig. 3.8 Setting up the DE Director

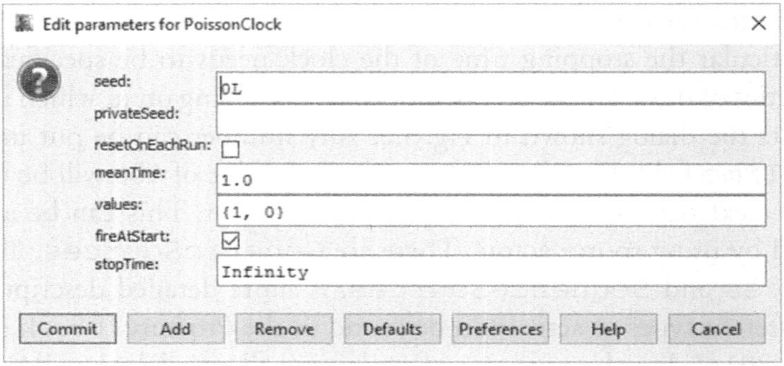

Fig. 3.9 Setting up the `PoissonClock`

to generate tokens, the default is `1, 0`, which implies that the clock will alternate between `1` and `0`. This list can be changed, but usually the case is that the output from a clock is somehow post-processed to create a meaningful quantity from it.

The `PoissonClock` default parameter for the seed is `0L`; this default value is interpreted by Ptolemy II as no seed specified, and the seed is then initialised to `System.currentTimeMillis() hashCode()` + (here `hashCode()` returns a code for the actor). It also needs to be noted that `seed` is a shared parameter, meaning that all random number actors share the same seed. If the user wants to specify a specific seed to be used by this particular actor that is private just to him and not shared, then he needs to enter a value in the `privateSeed` field, which will override the usual behaviour of the shared seed parameter. If the clock is connected without any modification to a `timedPlotter`, which is a `sink` of the type `timedSinks`, we will be able to visualise the output of the `PoissonClock` using a model such as the one presented in Fig. 3.10. The output of the `timedPlotter` is shown in Fig. 3.11.

It can be seen that the default format of the `timedPlotter` makes interpretation difficult in the present instance. To avoid such problems, the reader is advised to edit the format of the `timedPlotter` as indicated in Fig. 3.13. The `timedPlotter` then changes as shown in Fig. 3.12.

The `timedPlotter` graph looks better. The reader should note how the model alternates between the values `1` and `0`. If the `values` list of

3.4 An Illustrative Example 57

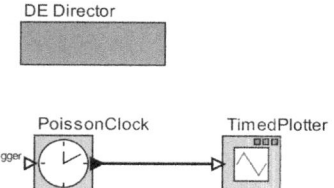

Fig. 3.10 `PoissonClock` verification

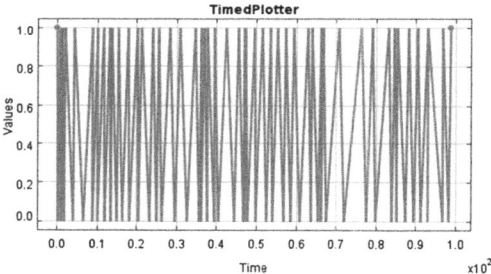

Fig. 3.11 Output of the `timedPlotter` actor for model of Fig. 3.10

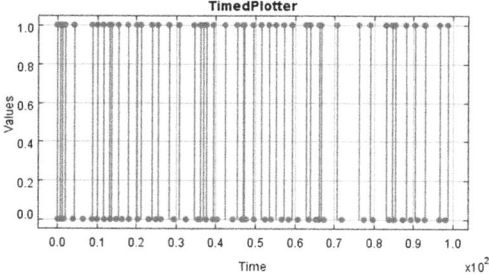

Fig. 3.12 Output of the `timedPlotter` actor for model of Fig. 3.10 after reformatting

the `PoissonClock` is changed now to 0,1,2,3, the output of the resulting `timedPlotter` can be seen in Fig. 3.14.

The next step is to introduce the `Ramp` actor. The actor operates by increasing its value according to the rule provided on its parameters. The default parameters are provided in Fig. 3.15. The model as it looks after

58	3 Discrete Event Simulation Concepts

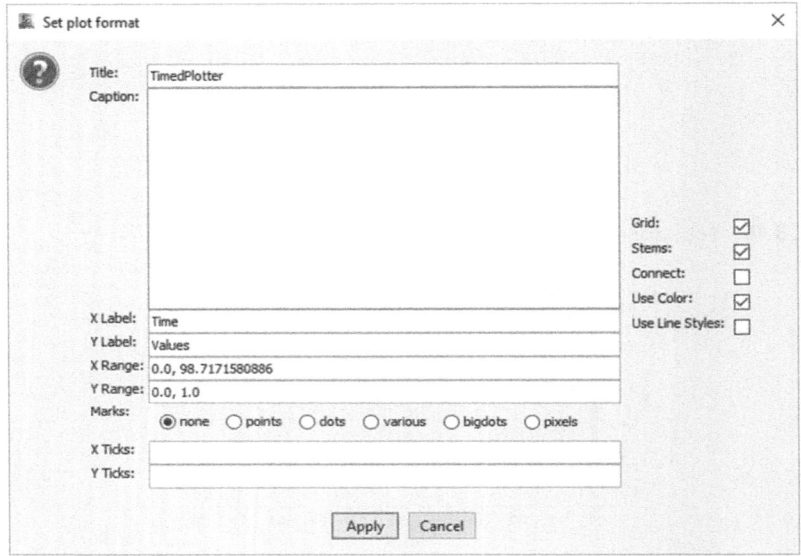

Fig. 3.13 Format editor for the `timedPlotter`

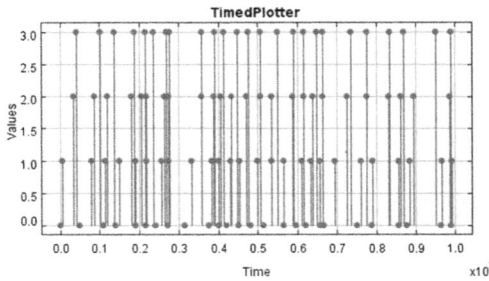

Fig. 3.14 Output of the `timedPlotter` actor for model of Fig. 3.10 with `values` equal to `0,1,2,3`

this step is presented in Fig. 3.16 and the output the model produces is presented in Fig. 3.17.

Regardless of the input, on each firing the `Ramp` actor will produce a token that will be one more than the previous (or whatever the `step` parameter is) starting at an `init` value.

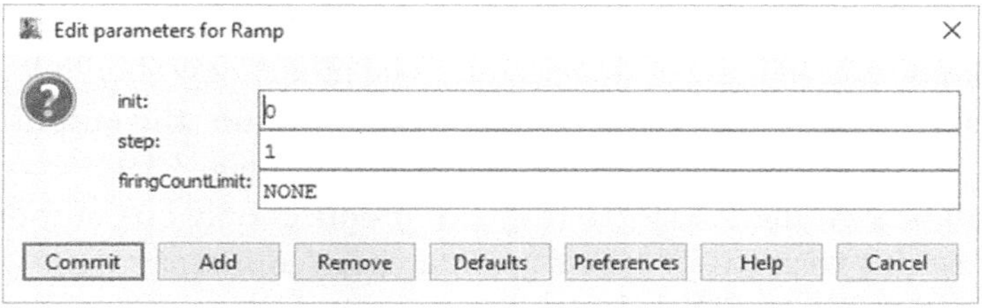

3.4 An Illustrative Example

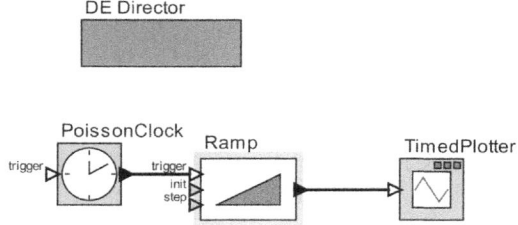

Fig. 3.15 Ramp actor default parameters

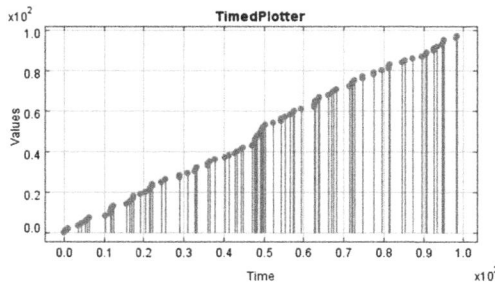

Fig. 3.16 Aspect of the model after Ramp actor is added

Fig. 3.17 Output of the model after Ramp actor is added

The next two actors that are added to the model are a Queue and a Server. These can be obtained from the DiscreteEvent folder (inside DomainSpecific). Then the output of the ramp needs to be sent to these two actors. For this reason, a relation is created (this can be done by clicking with the left button while holding the Ctrl key). The model also needs an additional timedPlotter actor. The first timedPlotter will be connected to the output port of both the

Queue and Server. The other timedPlotter will be connected to the size port of each one of these actors. The model now should look like the model presented in Fig. 3.18. If the model is run, the output of the two timedPlotters will look as illustrated in Fig. 3.19.

It can be seen that timedPlotter produces only one output. This is because this timedPlotter actor has been connected to the Queue and Server outputs, but the Queue actor is not being fired to produce output. The output of timedPlotter2 shows that the queue size of the Queue grows, while the queue size of the Server actor remains under control. To fix this, there are some amendments needed in the model. First, the service rate of the Server actor must be fixed, and second, a proper firing actor needs to be introduced to activate the Queue actor. In order to fire the Queue actor a Poisson clock needs to be connected to the trigger input port of the Queue actor. The Server actor service time will be generated using an exponential distribution. The

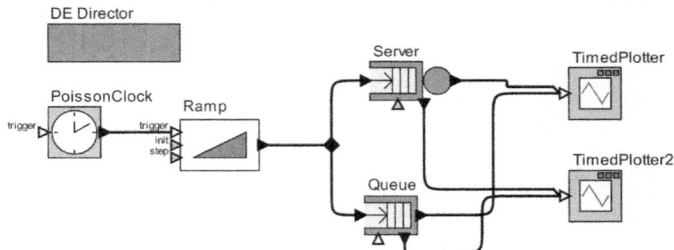

Fig. 3.18 Model after addition of the Queue and Server actors

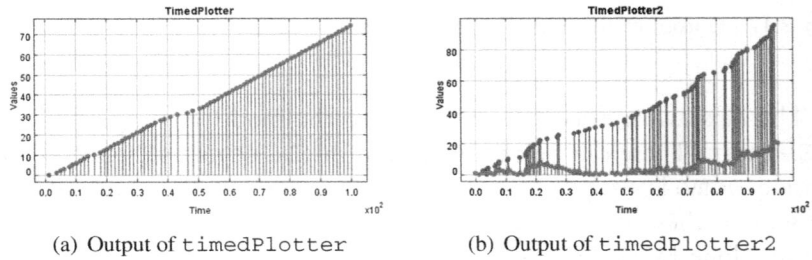

(a) Output of timedPlotter   (b) Output of timedPlotter2

Fig. 3.19 Output of the two timedPlotter actors

model now should look as the model presented in Fig. 3.20, the output is presented in Fig. 3.21.

It is important to observe the rates used for the additional actors were 0.5 for the PoissonClock and 2 for the ColtExponential lamda parameter (as the λ parameter is the inverse of the average service time).

3.5 Summary

This chapter has introduced the basic elements of simulation, and in particular has centred the discussion on DES models. The chapter does not pretend to be an exhaustive introduction to the topic, and the reader is advised to consult one of the several excellent textbooks available on the market.

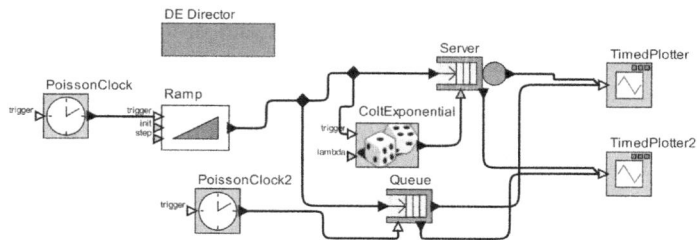

Fig. 3.20 Model after addition of the PoissonClock and ExponentialDistribution actors

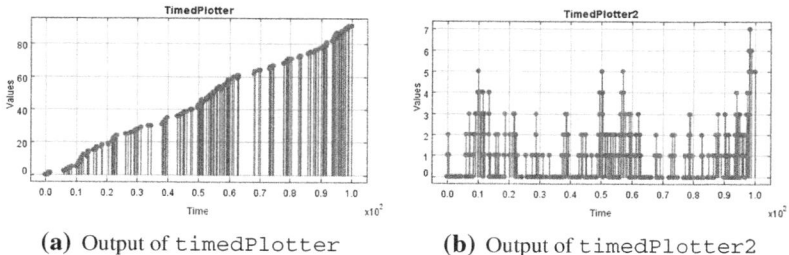

(a) Output of timedPlotter   (b) Output of timedPlotter2

Fig. 3.21 Output of the two timedPlotter actors

The software that will be used in the rest of the book was introduced in this chapter, and an example to illustrate the tool was analysed in some level of detail. The next chapter will guide the reader through the construction of the building blocks of a model for the Thai rubber industry supply chain, and some basic tests will be conducted to check the logical consistency of these elements.

3.6 Note Regarding Ptolemy II Images Used in the Text

The Ptolemy II images and screen-shots used in the chapter are subject to the following license:

> Copyright © 1990–2017 The Regents of the University of California. All rights reserved.
>
> Permission is hereby granted, without written agreement and without license or royalty fees, to use, copy, modify, and distribute this software and its documentation for any purpose, provided that the above copyright notice and the following two paragraphs appear in all copies of this software.
>
> IN NO EVENT SHALL THE UNIVERSITY OF CALIFORNIA BE LIABLE TO ANY PARTY FOR DIRECT, INDIRECT, SPECIAL, INCIDENTAL, OR CONSEQUENTIAL DAMAGES ARISING OUT OF THE USE OF THIS SOFTWARE AND ITS DOCUMENTATION, EVEN IF THE UNIVERSITY OF CALIFORNIA HAS BEEN ADVISED OF THE POSSIBILITY OF SUCH DAMAGE.
>
> THE UNIVERSITY OF CALIFORNIA SPECIFICALLY DISCLAIMS ANY WARRANTIES, INCLUDING, BUT NOT LIMITED TO, THE IMPLIED WARRANTIES OF MERCHANTABILITY AND FITNESS FOR A PARTICULAR PURPOSE. THE SOFTWARE PROVIDED HEREUNDER IS ON AN "AS IS" BASIS, AND THE UNIVERSITY OF CALIFORNIA HAS NO OBLIGATION TO PROVIDE MAINTENANCE, SUPPORT, UPDATES, ENHANCEMENTS, OR MODIFICATIONS.

Notes

1. Typical assumptions used in Engineering are that the system is described by a set of linear equations or that the underlying model satisfies a given condition (e.g. the matrix is definite-positive).
2. At this point it is assumed that the reader has followed the instructions provided in Chap. 1, installed Ptolemy II on his/her computer and has allowed enough time to explore the system. Various details will be provided later in the book, which for the novice could be overwhelming if there is not enough familiarity with the tool. In the case of any questions relating to Ptolemy II, the reader is encouraged to consult the software documentation.

References

Heidergott, B., Vázquez-Abad, F. J., Pflug, G., & Farenhorst-Yuan, T. (2010). Gradient estimation for discrete-event systems by measure-valued differentiation. *ACM Transactions on Modeling and Computer Simulation, 20*(1), 5:1–5:28.

Jerry, B., Carson, J. S., Nelson, B. L., & Nicol, D. M. (2010). *Discrete-event system simulation* (5th ed.). Pearson.

Law, A. (2015). *Simulation modeling and analysis* (5th ed.). McGraw Hill.

Lawrence, M. L., & Stephen, K. P. (2006). *Discrete-event simulation: A first course.* Pearson Prentice Hall.

Merriam-Webster. (2016). Merriam-webster's learner's dictionary. Retrieved October 31, 2016 from, http://www.merriam-webster.com/dictionary/.

Ptolemaeus, C. (2014). *System design, modeling, and simulation using ptolemy II.* Ptolemy.org.

4
A Hands-On Development of a Discrete Event Simulation Model for the Thai Rubber Industry

Abstract This chapter presents a discrete event simulation (DES) model which is built step by step, using Ptolemy II. The elements of the model are explained. After the design decisions have been made, the elements are implemented and integrated into the model being built. In each step of the process, checks and control measures are put in place in order to guarantee the validity of the model in a way that resembles unit testing in programming.

Keywords Discrete event simulation · Rubber industry · Ptolemy II

This chapter will present a discrete event simulation (DES) model that will be built step by step using the Ptolemy II software tool. The elements of the model will be explained and the overall model will be then discussed.

The construction of a model should follow a sequence of steps which this chapter will illustrate. A general idea of the type of the model and the units that will be used (time, distance, etc.) is given before implementation is discussed. After the design decisions have been taken, the elements can be implemented using the software, and any disparate elements can then be integrated. At each step of the process, checks and control measures are put in place to guarantee the validity of the model in a way that resembles unit testing in programming. The details are further explained in later sections of the chapter.

4.1 Planning Work

The objective of the model under discussion is to represent a supply chain. As described in previous chapters, the supply chain for rubber in Thailand has very distinctive elements, and each can be associated with a different "step" in the process of transforming rubber primary products into products ready to be used for further production of finished products.

The model described here consists of stages, with the earlier stages corresponding to raw materials which are moved, stored and processed in the supply chain. Figure 4.1 describes the value chain for the production of rubber products. The model has three clearly identified stages: farmers, intermediate distribution and factories. It has been decided that due to its complexity, the market will not be modelled and will be assumed as given and represented by an external demand for rubber products. This external demand for secondary product will be imposed on factories, which in turn will require primary raw materials (via a derived demand) from distribution centres.

In Fig. 4.1, below each component of the supply chain there is a symbol representing the number of each one. As a general rule, $M \geq N \geq O \geq P$. The farmers will vary in number from province to province and also in their mode of production; however, the farmers are more numerous than the intermediate distribution centres. The same argument applies to factories, and in general, the number external markets for rubber should be less than the number of factories.

The model in mind for the supply chain is one in which farmers push production to the intermediate distribution centres, which in turn pas product on to factories who respond by satisfying external demand (from the markets). These two elements define the ways the actor (farmer,

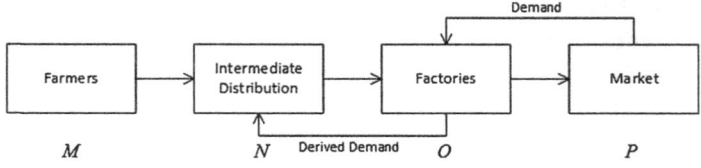

Fig. 4.1 A schematic representation of the supply chain

distribution centre and factory) not only connect but also interconnect. Some actors will need to create tokens that will be assigned physical properties (e.g. quantity). Other actors will need to store tokens, or at least account for the production of the farmers (distribution centres). Finally, there are the factories that will create tokens to extract material from the distribution centres. This last point conveys an implicit assumption that factories do not have or need large stockpiles of material.

It is assumed that the transactions of tokens between actors happens once a day. This assumption is made to simplify the creation of tokens and also to avoid overpopulating the future events that can be summarised in daily transactions. Clearly, these assumptions could be removed, but in order to keep the model simple, they create a more realistic simulation of the material flow in the supply chain for future work.

Now that the basic concept of the type of model is given, its implementation will be carried out in Ptolemy II for each of its elements. The descriptions in the following sections proceed from "left" to "right" in the supply chain. With respect to order, movement would be expected from producer to intermediates, to final arrival at the factories, which produce the final output to be transported using outbound logistics.

4.2   The Director

The first step in the construction of the simulation model is to add *the director* for the simulation. This can be performed by clicking on the *Directors library* on the left-hand side, and from there dragging and dropping into the canvas area the appropriate *director*. In our case we need the DE Director. With this, the canvas now appears as in Fig. 4.2. It should be noted that by double-clicking the director, the properties can be accessed, and they must be set before running the simulation. As the unit for time will be minutes, the amount of time must be set for the simulation. Simple calculations show that in a day, there are 24 * 60 = 1440 min. Therefore, with 30 days in a month, the time would be 1440 * 30 = 43200 min. Consistently, the value 43200 must be put into the stopTime field. The user then clicks on Commit in the same dialog box as shown in Fig. 4.3.

In order to check that the model runs effectively for the amount of time being specified, three actors are added to it: a `Discrete Clock`, a `Ramp` and a `Monitor Value` (there are no changes needed in the properties of these actors). They are linked, and then the model is run, and the value registered on the `Monitor Value` actor, which should be 43200, may be observed. The `Discrete Clock` ticks once per unit of time, and as the time has been set to 43200, it will "tick" up to the 43200 mark. The `Ramp` actor will increase by one unit each time it is triggered, which will in turn update the `Monitor Value` actor, which will change accordingly until the simulation is finished. An illustration of the configuration of the model and its output is provided in Fig. 4.4. The output confirms that the model runs as it should for the desired time. Obtaining a different result could signify potential problems. In this case, it would be necessary to review the different actors in the model to check that parameters are appropriate.

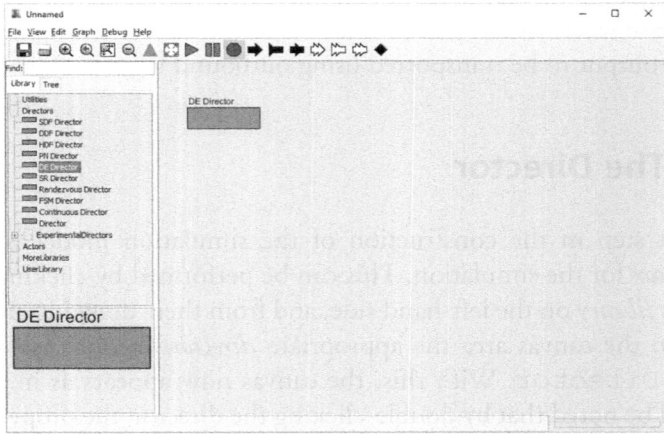

Fig. 4.2 A Vergil graph editor with a DE Director

4.3 Adding a Source Clock

For the simulation to proceed, tokens are created within the system. Tokens may represent interactions with the external system (e.g. customer arrival) or signals created by sources in the system. For example, the farms in this case create liquid and solid rubber primary products every day, which later flow into the supply chain. In order to show the tokens coming into the system, a clock is added by dragging it from the left-hand side of the Library panel. The modelling choice that should be adopted is that every province will produce as many signals as farmers, with the size of those farmers following a power law (as discussed previously). However, this approach could create far too many tokens in the system. This would affect the execution speed of the model, as the future event list would grow substantially and produce a corresponding overload, so this decision needs to be carefully weighed. At this point in the construction of the model, the clock is set to create one event per day, which triggers information about each province and the creation of product loads in the system. The clock that is added then is a Discrete Clock (which is named the MasterClock), and it produces a token once every 1440 units of time, i.e. once a day (see Fig. 4.5). A test similar to a test performed previously (and represented in Fig. 4.4) was performed to check that 30 tokens were produced.

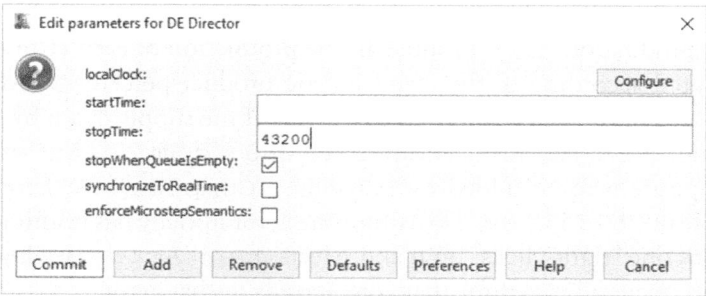

Fig. 4.3 Setting the time in the DE Director actor

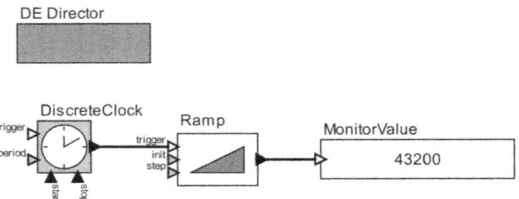

Fig. 4.4 Setting the time in the DE Director actor

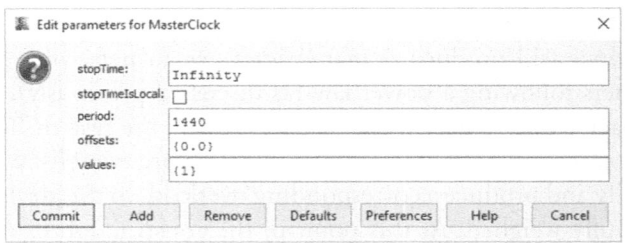

Fig. 4.5 Setting the parameters of the clock

4.4 Modelling the Production of Each Province

In order to model the aggregated production within each province, a set of *farmers* will be created that follow a power law distribution for size, and from here (based on a productivity parameter per unit of land), a *farmer* production unit is produced. The production of each farm is then distributed according to the characteristic product percentages for each province. This is then transferred to the rest of the supply chain by means of output ports. All this is encapsulated into a `composite actor`. A composite actor is added by dragging a `CompositeActor` actor available in `Utilities`. The composite actor initially has nothing on it; however, one "right-clicks" on it and selects `Open Actor` (see Fig. 4.6), and this opens a new editor that corresponds to that actor.

4.4 Modelling the Production of Each Province

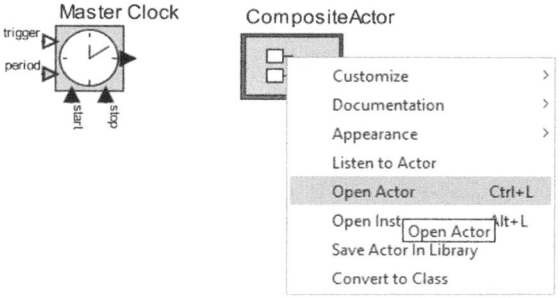

Fig. 4.6 After dragging the composite actor, it is opened by right-clicking and showing the contextual menu

4.4.1 Using Power Laws to Model Production

To understand the distribution that needs to be used, a small mathematical *intermezzo* is employed. Adopting a very "loose" definition, it can be said that a random variable X follows a power law probability distribution if its probability density function can be expressed as $f(x) = C \cdot x^{-\alpha}$ for $x \geq x_{min}$ and $\alpha \geq 1$, where the normalisation constant $C = (\alpha - 1)x_{min}^{\alpha-1}$ is obtained by applying the usual condition $\int_{-\infty}^{\infty} f(x)dx = 1$. This will imply the following:

$$\int_{-\infty}^{\infty} f(x)dx = \int_{-\infty}^{\infty} Cx^{-\alpha} dx = C \int_{x_{min}}^{\infty} x^{-\alpha} dx \text{ because } x \geq x_{min}$$

$$= C \left[\frac{1}{1-\alpha} x^{1-\alpha} \right]_{x=x_{min}}^{x=\infty} = -\frac{C}{1-\alpha} x_{min}^{(1-\alpha)} = \frac{C}{\alpha-1} x_{min}^{(1-\alpha)}$$

imposing the condition $\frac{C}{\alpha-1} x_{min}^{(1-\alpha)} = 1$, yields the desired result.

One way of simulating a random variable following a power law with power α and with values restricted to the interval $[x_0, x_1]$ is to apply the following formula to uniformly distributed numbers y:

$$[(x_1^{(1-\alpha)} - x_0^{(1-\alpha)}) \cdot y + x_0^{(1-\alpha)}]^{(1/(1-\alpha))} \tag{4.1}$$

In Ptolemy II this can be achieved by using a setup similar to that of Fig. 4.7. The first actor is a Uniform random number generator and

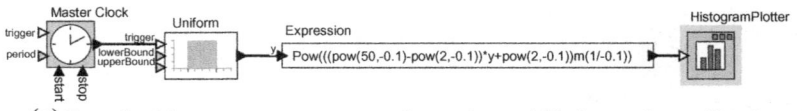

(a) Example of the setup to generate a power law random variable for $x_0 = 2$, $x_1 = 50$ and $\alpha = 1.1$

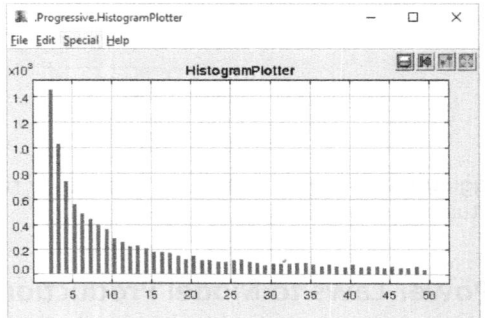

(b) Histogram created after 10,000 numbers were generated for $x_0 = 2$, $x_1 = 50$ and $\alpha = 1.1$

Fig. 4.7 Example of a bounded power law random variable generator in Ptolemy II

is available in the `Random` folder in the actor library. From there, `Uniform` needs to be selected and the parameters for this actor need to be `lowerBound = 0.0` and `upperBound = 1.0`.

4.4.2 Adding a Composite Actor

The composite actor must communicate with the rest of the model, so ports need to be added to it. These ports provide input to the composite actor (tokens are passed to it) and they provide output for the composite actor to communicate with the other actors. The menu for the ports can be accessed by right-clicking on the composite actor and then selecting `Customize` and from there, selecting `Ports` (Fig. 4.8).

This last should open the dialog box, as in Fig. 4.9, where the user can add and edit ports. Ports can be specified as input or output ports; they can be named, and their appearance modified. The dialog is fairly

4.4 Modelling the Production of Each Province

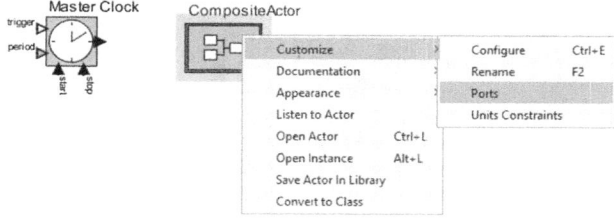

Fig. 4.8 The interface to define the ports can be accessed by right-clicking the composite actor and selecting `Customise` and then `Ports`

straightforward. Click on `add` to add a port, then type in the name and select the type of port.

4.4.3 The Model

The model built for each province (inside the composite actor) is presented in Fig. 4.10. It consists of the following elements:

- Parameters, which allow for values to be set and used by other actors. In the particular case of the model created to represent farmer production in a province, the following parameters were implemented:
    1. LX Proportion: The proportion of the total production of the province that will be destined to become *field latex*. This is initially assumed to be a fixed percentage, but a different option (such as a variable in time) could be modelled.

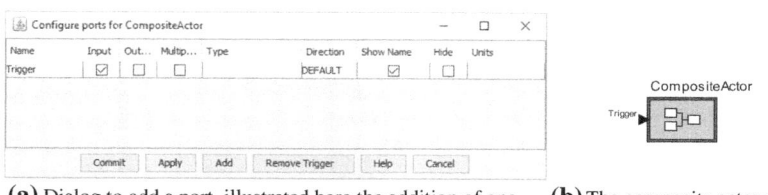

(**a**) Dialog to add a port, illustrated here the addition of one input port

(**b**) The composite actor with a named input port

Fig. 4.9 Adding a port to a composite actor

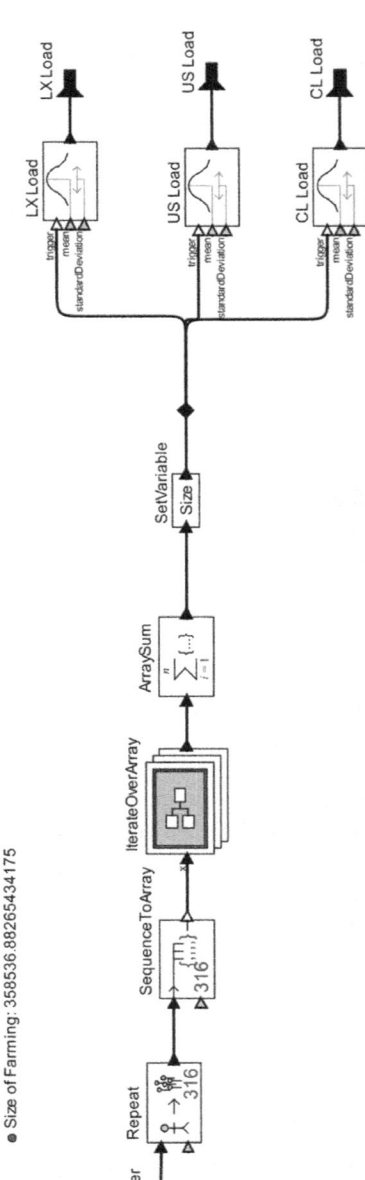

Fig. 4.10 The template model for each province

4.4 Modelling the Production of Each Province

2. US Proportion: The proportion of the total production of the province that will be destined to become *unsmoked sheet*. This is initially assumed to be a fixed percentage, but a different option (such as variable in time) could be modelled.
3. CL Proportion: The proportion of the total production of the province that will be destined to become *Cup-Lump*. This is initially assumed to be a fixed percentage, but a different option (such as variable in time) could be modelled.
4. Productivity per rai[1] could also be used. An average value of production per rai is assumed for each province. Please note that as the farmers are generated following a power law, there is already an element of uncertainty being introduced into the model.
5. Number of Farmers: A property of each province.
6. Size of Farming: This variable is set up to provide an initial value to the variable `Size` which is declared after the array of farmers is processed. The parameter is named the same as the `SetVariable` variable name, which allows for this variable name to be initialised. This initialisation will be overwritten when the `SetVariable` actor assigns a value to the `Size` name.

- Expressions, which act on an input to produce an output using mathematical formulae. It is important to note that if using variables, those variables need to be initialised prior to use or this will raise an error. That is the reason there is a parameter named `Size` that is assigned to a variable actor but also initialised via a parameter.
- A variable, which is a placeholder for a value that is dynamically set and that allows the representation of a quantity that changes over time.
- Random number generators (RNGs). There are two types used in this particular model: a `Gaussian` and `Uniform` RNG actors. The `Uniform` RNG is used in conjunction with an `Expression` actor to create a power law random variable sampler (as discussed previously) and the `Gaussian` RNGs are used to generate the different productions at each period of time.
- A `Repeat` actor that allows multiple generation of tokens. For the particular case of the model under examination, the repeat actor is used to simulate the production of different farmers within a province. In

order to achieve this, several tokens (as many as the number of farmers) are generated and then converted to an array, which in turn is passed to another actor that will transform it into an array.

- A `SequenceToArray` actor that takes the output of a repeat actor (a `sequence`) and transforms it to an `array` type that can be further processed.
- An `IterateOverArray` actor that allows consumption of the signals created by the `Repeat` actor and iterates over each element and applies the same logic to each element.
- An `ArraySum` actor that takes an array as input and produces the sum of the array as output. In the particular case of the model we have created, it takes the different amounts of production of each farmer and accumulates all of them to provide the production of the province.

The most complicated piece of modelling inside the composite actor is the transformation of the `Repeat` tokens into an array and then the post-processing of it to produce a usable number for the model. In order to understand why this is necessary and important, consider the example described in Fig. 4.11. This simple example generates 20 repetitions of a token, which are then used to activate a Uniform RNG, which will in turn produce tokens. When the model is executed, the tokens are produced all at the same time, as shown in Fig. 4.12.

This type of output is not suitable for further processing. If we assign it to a `SetVariable` actor, then essentially it will take only one of the variables of the set and assign it to the variable the same number of times as the `Repeat` actor establishes. This means that the model will lose all the information relative to the distribution of farm sizes and will keep only one, which actually defeats the purpose of using the `Repeat` actor in the first place. If the reader is not sufficiently convinced of this, then they could replace the `TimedPlotter` of Fig. 4.11 with a `Display`

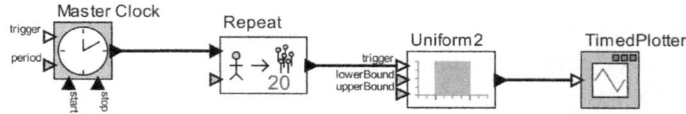

Fig. 4.11 Iterate Over Array actor implementation logic

4.4 Modelling the Production of Each Province

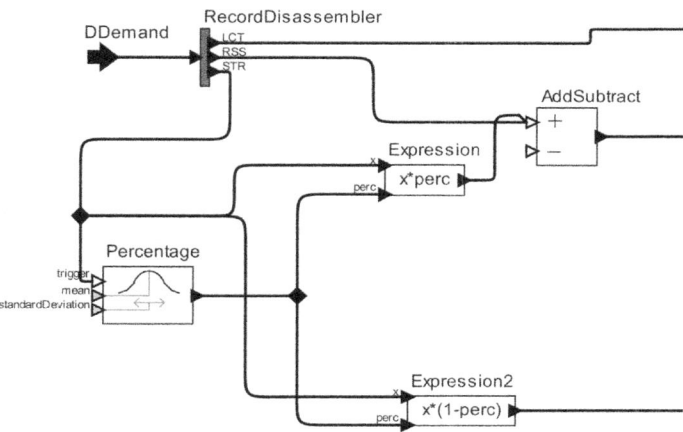

Fig. 4.12 Output of the execution of the repeat model

actor and then insert a SetVariable[2] actor in between the Uniform and Display actors. If a SequenceToArray actor is used then the Display actor outputs the following:

```
{0.1319850361535, 0.1800388690664, 0.6382923014756,
0.5470961514137, 0.2019881258288, 0.3931982322748,
0.7292410784998, 0.1060869131811, 0.5850431889925,
0.5301466034059, 0.9617861267974, 0.0795474089314,
0.4349752914044, 0.7923685898959, 0.2001147387552,
0.4239088850845, 0.2117744066349, 0.0745656622805,
0.9167869533224, 0.8916284259035}
```

The implementation of the logic associated with the IterateOver Array actor is given in Fig. 4.13. It needs to be noted that an Iterate OverArray actor requires the provision of a Director actor. According to Ptolemy's documentation, "The Synchronous Dataflow (SDF) domain supports the efficient execution of Dataflow graphs that lack control structures", and one of the examples available in Ptolemy's documentation uses this actor for the same purposes needed by our model, so it will be the director that will be used in the model.

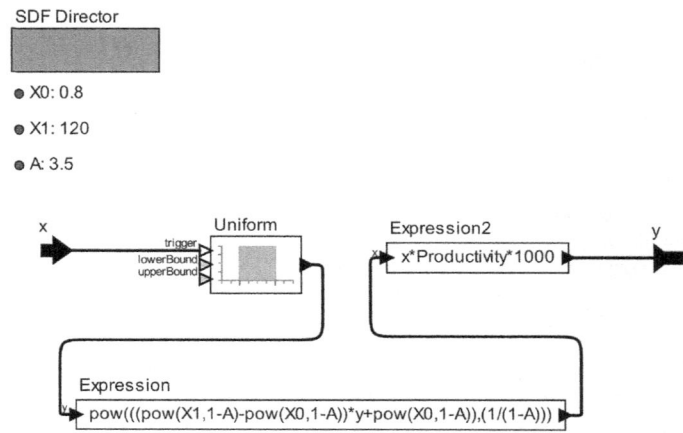

Fig. 4.13 Iterate Over Array actor implementation logic

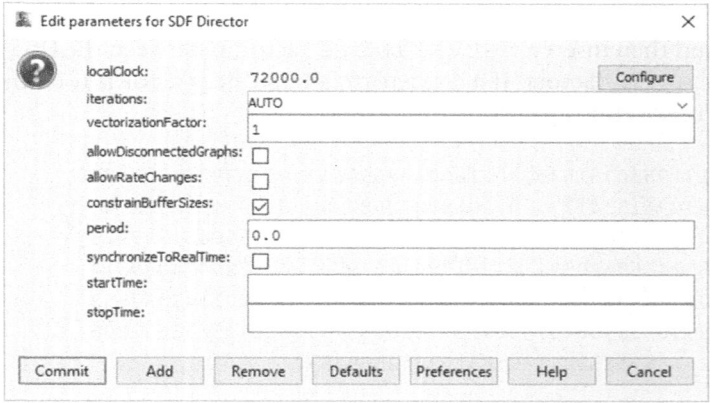

Fig. 4.14 The SDF Director for the Iterate Over Array actor

The parameters used in the SDF Director were the default ones and are represented in Fig. 4.14. It also needs to be noted that the IterateOverArray actor specified three parameters X0, X1 and A that represent, respectively, the lower bound, the upper bound and α for a bounded power law distribution. Two ports were also added to the

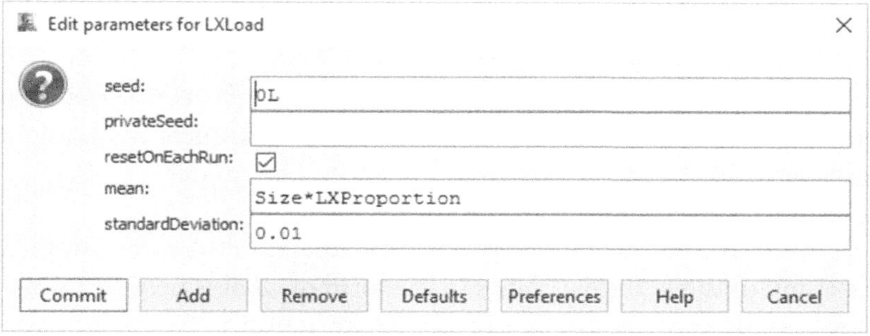

4.4 Modelling the Production of Each Province

Fig. 4.15 Example of parameter setting for one `Gaussian` actor

`IterateOverArray` actor: an input port named `x` and an output port named `y`. The input port receives a token and immediately creates a `Uniform` random number (i.e. in the interval [0, 1]). Applying the transformation discussed in Eq. 4.1, this output, which represents the size of a particular farm, is then scaled to take into account the productivity of the farm. The `Productivity` name has already been defined in the actor that contains the current actor. Therefore, the name is available and well defined and can be used (the expression will not raise an error). In fact, if the reader attempts to change the name of the parameter in the level that contains the `IterateOverArray` actor, then an error will be raised.

The last element missing from this model is the generation of different products. This is achieved using a `Gaussian` actor. This actor receives the Size value and distributes it according to the percentage defined in the `LXProportion`, `USProportion` and `CLProportion` parameters. The process is simple, as the names for those proportions are already defined and are model parameters. They can be used in expressions, as Fig. 4.15 suggests.

4.5 Modelling the Distribution Centres

The distribution centres are the composite actors that receive the output from the provinces and stock the different raw products that later will be delivered to the factories (upon demand). The model is presented in Fig. 4.16.

Figure 4.16 essentially shows two main elements of the composite actor: inflow and outflow. Inflow comes in the form of what is being sent by the farmers, outflow is what is taken due to the action of external demand. These two elements are examined later in more detail. Additionally, a simple parameter called `Capacity` is added to limit the storage capacity of the distribution centres. This parameter changes from province to province.

4.5.1 Inflow to the Distribution Centres

The inflow to the distribution centres is incorporated into the composite actor by using the input ports `LXLoad`, `USLoad` and `CLLoad`. These

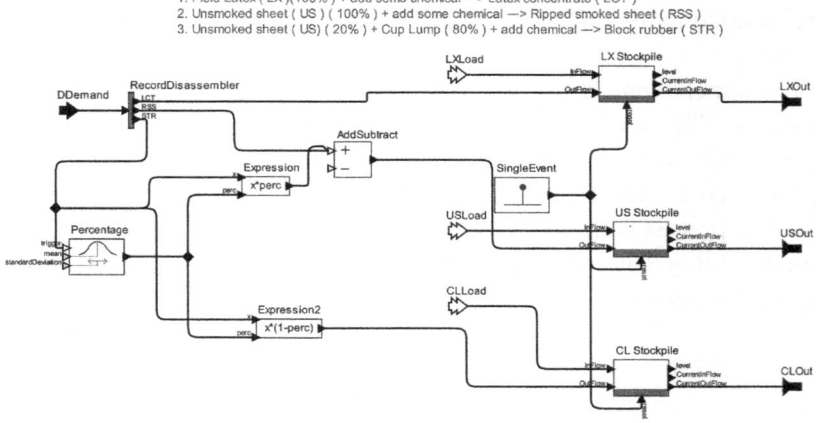

Fig. 4.16 Example of a distribution centre

4.5 Modelling the Distribution Centres

input ports are linked to the output ports of the `Province Load Generator` composite actor that we have already described.

The different loads are then "stored" into what has been termed *Stockpiles* (`LX Stockpile`, `US Stockpile` and `CL Stockpile`). It is noteworthy that this "storage" is a composite model of its own.[3] The `Stockpile` composite actor can be described by means of Fig. 4.17.

The `Stockpile` model has three inputs and three outputs (not all of them are required as explained later in the book). The three input ports are `InFlow`, `OutFlow` and `reset`. What the model tries to achieve logically is very simple: it adds to the running balance if a token comes from the `InFlow` and subtracts from it if it comes from the `OutFlow`. The composite actor also has three parameters: `capacity`, `initialLevel` and `actualLevel`, which are used in expressions for the logic of the inflow and outflow.

When a token is received in the `InFlow` port it is then evaluated by an `Expression` actor containing

`(actualLevel + x < capacity) ? x : 0`

This expression is a conditional statement that compares the `actualLevel + x` quantity against the `capacity` parameter.[4] Note that x in the expression represents the token entering the expression via the input port x. If the comparison is true (i.e. if `(actualLevel + x < capacity)`), then the expression will output x; otherwise it will return 0. The output of the `Expression3` actor is then connected to an `AddSubtract` actor in the + input port. This actor output is, in turn, connected to both an `Accumulator`, which adds up numbers for the total, and an output port (`CurrentInFlow`) that could be used to examine the incoming flow.

Equally, when a token is received in the `OutFlow` input port, it can be evaluated by the expression

`(actualLevel - OutFlow > 0) ? OutFlow : actualLevel`

i.e. if the desired amount of rubber from the stockpile can be served by the current level then it extracts all this. Otherwise it will simply extract whatever is left (represented by `actualLevel`). The output of this expression is connected to the – port of the `AddSubtract` actor which, as mentioned previously, is then connected to both the `Accumulator` and

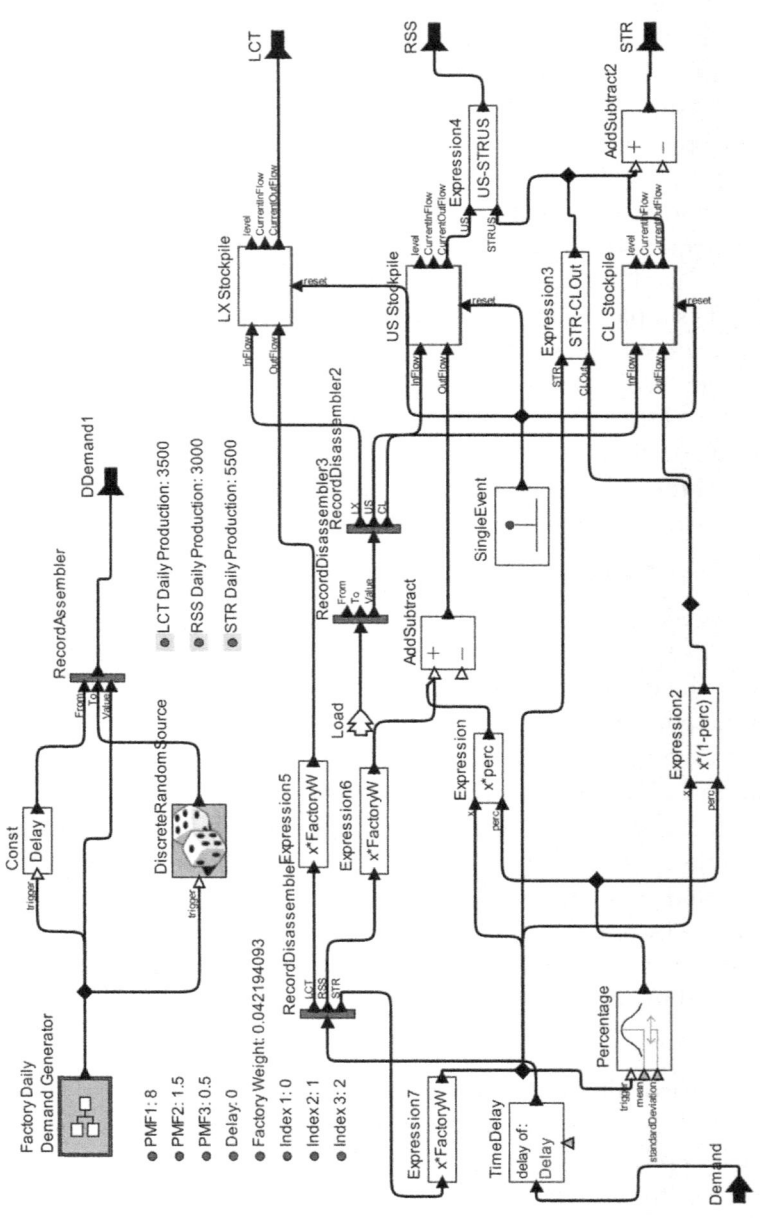

Fig. 4.17 Implementation in Ptolemy II of a `Stockpile` composite actor

4.5 Modelling the Distribution Centres

to an output port (`CurrentOutFlow`) that could be used to examine the outgoing flow. Finally, the output of the accumulator is connected to an output port (`level`) and it is also used to set the variable `actualLevel`. Please note here that the `reset` input port is used by an external one-off event (fired at the start of the simulation) that sets the `actualLevel` variable to zero.

A final point regarding the animated custom icon of the composite actor: It can be edited by navigating to the context menu (right-clicking on the composite actor), as described in Fig. 4.18. From there, a special editor may be accessed, as shown in Fig. 4.8. Here can be seen two rectangles, one on a *blue* background (the user can choose whichever colour is desired). Over the blue rectangle is a white one, initially of the same size as the background rectangle. As the simulation progresses, there is more material in the stockpile, and the white rectangle should shrink (i.e. reduce in height) in order to reveal the blue rectangle (Fig. 4.19). This process is controlled by using the parameters presented in Fig. 4.20. Note the use of the parameters that define the `actualLevel` and the `capacity` of the stockpile and that the rectangles are initially of size 60 by 40 (width and height respectively).

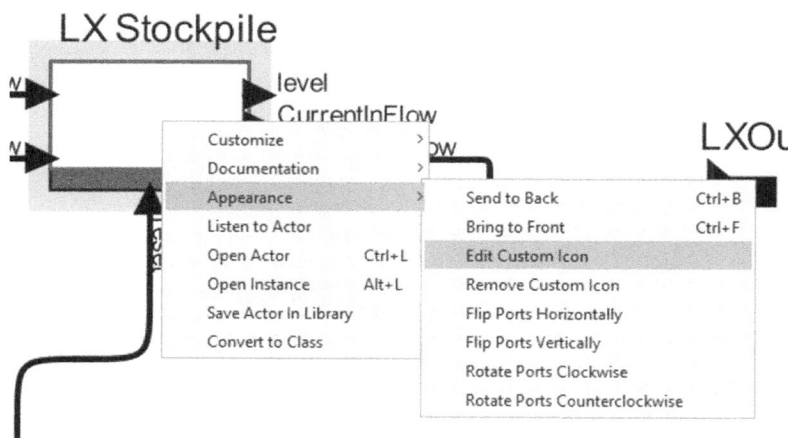

Fig. 4.18 Accessing the Custom Icon menu `Right-Click -> Appearance -> Edit Custom Icon`

As long as the levels are changing, the height of the white rectangle will be updated to reflect the current level. This will then animate the current level to go up or down depending on the inflows and outflows. This solution is also based on one of the examples available in Ptolemy's documentation, which is highly recommended as an authoritative reference.

4.5.2 OutFlow from the Distribution Centres

The outflow from the distribution centres is derived from the external demand and passed onto the distribution centres by the factories by using the `DDemand` input port, which connects to an output port from factories, as detailed later in the book. The mechanism to deal with this is similar in both cases as explained below.

The mechanism by which the external demand is produced is described later in the text. It is important to note that the demand is generated as a `Record`, which is one of the data structures that can be handled by Ptolemy II. A typical example of a record is given below:

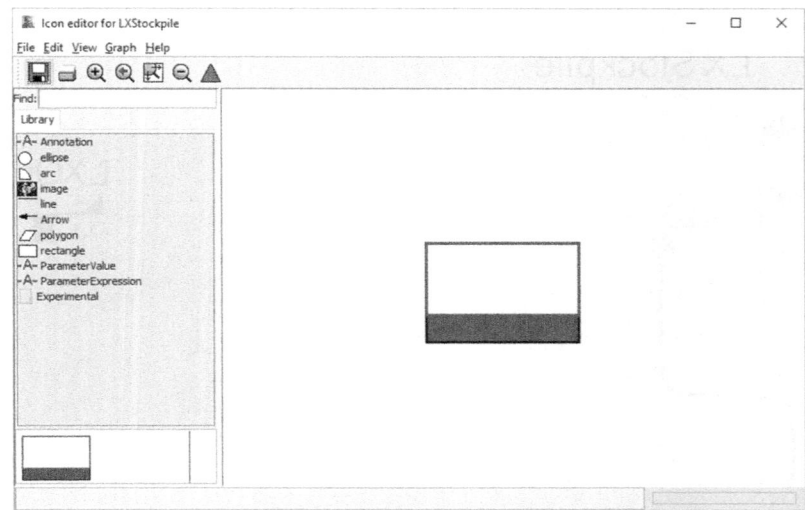

Fig. 4.19 Custom Icon editor

4.5 Modelling the Distribution Centres

```
{LCT = 3494.8971736210, RSS = 2996.2089978531,
STR = 5487.818056020}
```

Obviously, this type of data is not suitable for "consumption" by most of the other actors. Nevertheless, it is possible to "decode" this record by using an actor called a `RecordDissasembler`, which is shown in Fig. 4.21. This essentially takes the record and breaks it into as many output ports as it has. It needs to be noted that the output ports must be named the same as the names of the fields of the record.

Once the record has been disassembled, it needs to be further processed in order to determine the quantities of primary rubber products needed. There are several secondary rubber products that can be created. However,

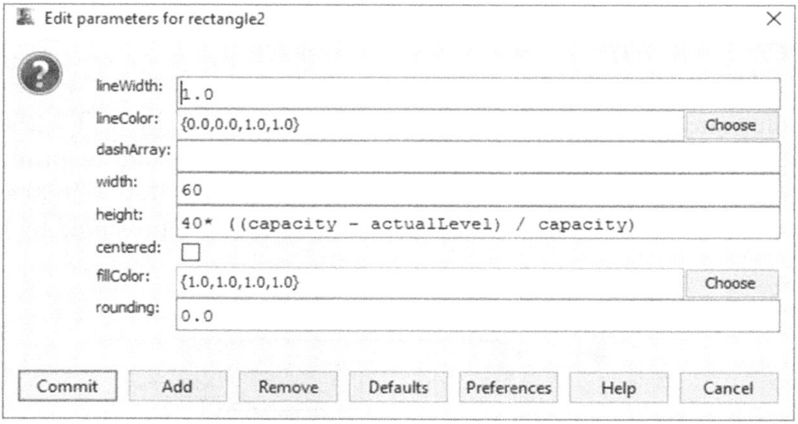

Fig. 4.20 Properties of the white rectangle for stockpile actor animation

for the purposes of building the first model it is simplified to only three: latex concentrate (LCT), ripped-smoked sheet (RSS) and block rubber (STR). The compositions of these secondary products can be described in the following way:

- LCT only requires LX
- RSS only requires US
- STR is composed of 20% US and 80% CL

For this reason, given a demand for LCT, RSS and STR, this will then create a derived demand for the primary products described by the following relationships:

- $LX = LCT$
- $US = RSS + 0.2 \cdot STR$
- $CL = 0.8 \cdot STR$

To illustrate the point, consider that the demand is given by LCT = 2500, RSS = 3200 and STR = 2000. The corresponding demand for primary rubber products will be then: LX = 2500, US = 3200 + 400 = 3600 and CL = 1600. This requirement for primary product is represented by the

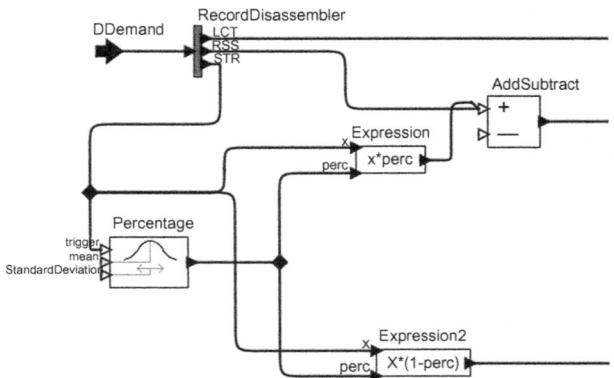

Fig. 4.21 Inputs to the distribution centre composite actor

operations produced on the tokens coming out of the output ports of the `RecordDissasembler`.

It can be observed that all of the LCT is connected to the `OutFlow` of the LX Stockpile composite actor, meaning that this demand will be discounted from that stockpile. All the material represented by the RSS token is added to a fraction of the STR token (20% of it), and this action is performed by an `AddSubtract` actor. Finally, the STR token is "split" into two quantities that represent roughly 20% and 80%. A decision was taken to not have this percentage as an exact number but to model it with a small amount of variability around it to account for imperfections in the process. Hence, the STR token is applied as a percentage (named `perc`) that is generated from a `Gaussian` distribution. The complement of `perc` is applied to the CL stockpile (this value can be calculated as (`1-perc`)). The split is made possible by the use of an `Expression` actor that has two inputs (`perc` and `x`). `x` receives the token representing STR and the expression is applied to each case.

4.6 Modelling the Factories

The factories are currently modelled in a manner somewhat similar to that of the distribution centres. The factories are modelled as composite actors with three inputs (LX, US and CL) and four outputs (LCT, RSS, STR and DDemand).

The `Factory` composite actor has a "decoding" mechanism that uses a `RecordDisassembler` actor and which works in the same way as previously explained. It also has Stockpile actors that are implemented in a similar way as previously explained (capacities may change). A visual representation of the `Factory` model can be seen in Fig. 4.22.

It is important to note the inclusion of a composite model named `Factory Daily Demand Generator`. This composite actor creates the daily demand of every `Factory` actor and depends on the parameters `LCT Daily Production`, `RSS Daily Production` and `STR Daily Production`. These parameters are input to each factory actor, and they are used by the `Factory Daily Demand Generator` composite actor. Inside this actor, the implementation is similar to that of the *external demand generator* (see Sect. 4.7 for additional details) with one substantial difference: the factory generator uses a `DiscreteClock` actor, while the external demand generator will use a `PoissonClock`. This model is currently very similar to the one presented for distribution centres, and will be kept that way for simplicity. There is a subtle difference between the two models in the way the output is calculated. If close attention is given to the two models, it can be seen that the stockpiles are for primary rubber products. For this reason, if the output from the stockpiles is used as the only number, what will be reported is the actual consumption from the stockpile but not the quantity of the secondary products produced. In order to properly account for the secondary products leaving factories two `Expression` actors are added:

- One to calculate `RSS` output which is equal to `US` minus the amount of `US` required in the production of `STR` (recall that the demand for `US` is modelled as $US = RSS + 0.2 \cdot STR$. The quantity removed to `US` has been called `STRUS` in one of the expressions and is calculated as per the second expression below.
- A second `Expression` to calculate the amount of `US` needed to produce `STR`. This quantity is calculated as 80% of the `STR` demand. However, as the percentage has a random element added to it, it is

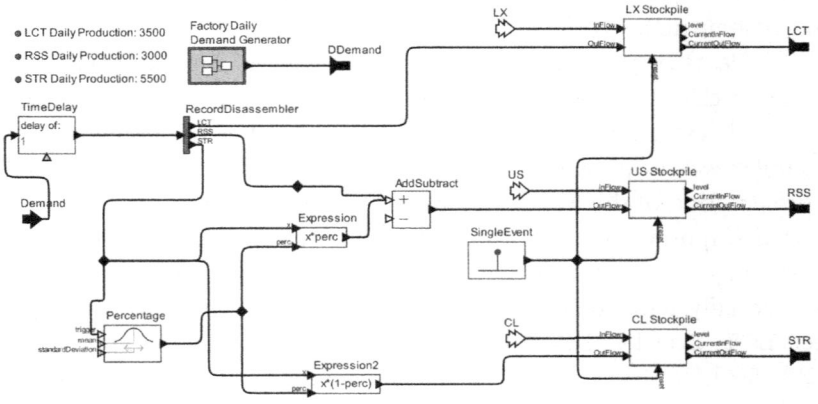

Fig. 4.22 Implementation of the `Factory` composite actor

then calculated with an expression of the form STR-CLOut, where CLOut is equal to STR*(1-perc) and perc following a normal distribution with a mean of around 20%.

The model could be expanded to consider inventory management policies for each factory, and eventually, *economic order quantity* optimisation that would then make every factory have a different frequency of ordering. However, it is believed that this topic is outside the bounds of this study and may be researched further at a later date.

4.7 Modelling External Demand

The last element in the model is one of *external demand*. The external demand is assumed to be known and is modelled as a random variable, which in each period will generate demand for each one of the three secondary products. Each one of the stochastic demands is modelled using a Gaussian actor. However, any other distribution type can be used. Essentially, the way to model this external demand is to take the total demand of a year and to assume that demand is evenly spread over the year (which may be an overreaching assumption).

The implementation of external demand can be seen in Fig. 4.23.

The only novel element that this part of the model brings is the use of a DiscreteTimeDelay actor whose purpose is to produce an output after the time specified as a parameter has elapsed. This gives the model an "opportunity" to deposit some material into the stockpiles before it is

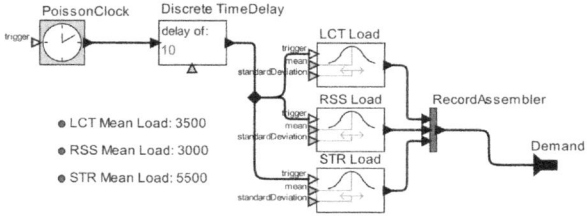

Fig. 4.23 Implementation of external demand in the model

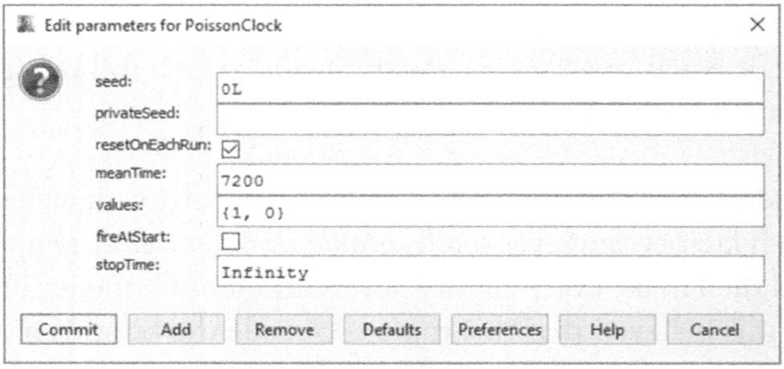

Fig. 4.24 Parameters of Poisson Clock

taken out. The other novel element is the use of a `RecordAssembler` that will take tokens at every input port and produce a record whose fields are named after the name used for the input ports.

It must be noted that the clock in this case is no longer a `Discrete Clock` but a `PoissonClock`. A `PoissonClock` generates events according to a Poisson distribution (i.e. one for which the times between events are distributed exponentially). For an example of the parameters that can be set on the Poisson Clock, please refer to Fig. 4.24. Please note that an option to *fire at start* can be activated or deactivated; we have chosen not to activate it.

4.7.1 Basic Validation of the Components of the Model

To understand the essential operation of the model and its consistency with regard to output, a basic example has been created below (see Fig. 4.25) to simulate a simple supply chain. From here, the outputs may be observed to check that the quantities are those that are expected.

For effective identification of the different elements of the chain (all of them modelled as composite actors), a change was made to the icons in a manner similar to that already described (i.e. using the icon editor). The model is run for a period of 43200 units of time, the Master Clock is set at intervals of 1440 units of time, the factory requests units of primary

4.7 Modelling External Demand

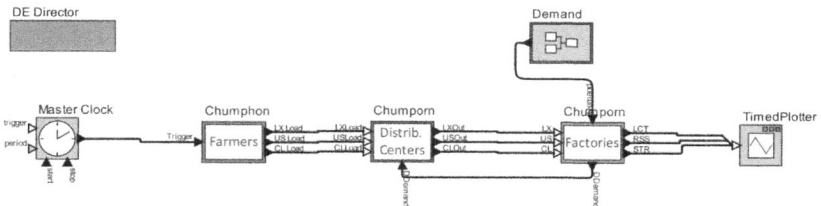

Fig. 4.25 Parameters of Poisson Clock

rubber products at intervals of 1440 units of time starting at 1450 (this one uses a delay of 10 units) that are tied to this daily production plan:

- LCT: 3500
- RSS: 3000
- STR: 5500

Finally, demand is configured with a clock, following a Poisson distribution, with the average time between events being 7200 units of time, and demand being the same as the daily production plan of the factory, each one modelled as a normal distribution on these means with a fixed standard deviation of 50 units. Figure 4.26 shows the results of one run. An appropriate match can be seen between the requirements of demand and the actual figures produced as outputs from the factories.

A small modification to the model can be performed in order to gain insight into what is happening inside the composite models. For this, two `TimedPlotter` actors are added to the factory composite model: one follows the levels of the stockpiles over time, and the other follows the amount of material extracted from each primary product stockpile. Figure 4.27 shows the model after these modifications were performed. Note that the `stockpile` composite actors have an output port for the `level`, and the `TimedPlotter` is connected to those. For the other graph, the input ports of the `stockpile` actors are connected to the `OutFlow`.

The output of the graphs is given in Figs. 4.28 and 4.29.

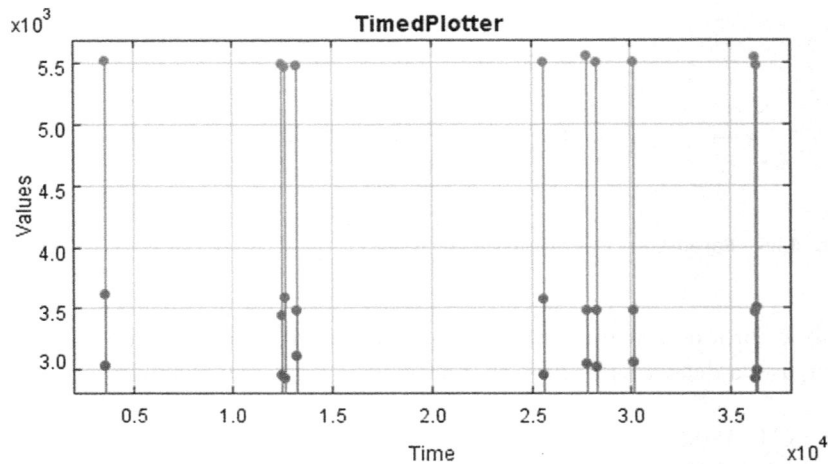

Fig. 4.26 Output of a simple simulation

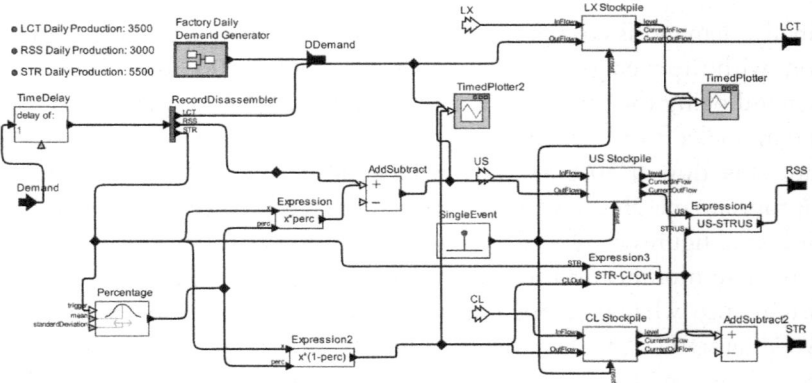

Fig. 4.27 Modification of the actor to add "listeners" for some quantities

4.8 Summary

In this chapter, the elements of a DES model for Thailand's rubber supply chain have been described. In the next chapter, a validation of the model will be presented, and the real configuration of the supply chain will be presented. The results generated by the model will be compared with his-

4.8 Summary

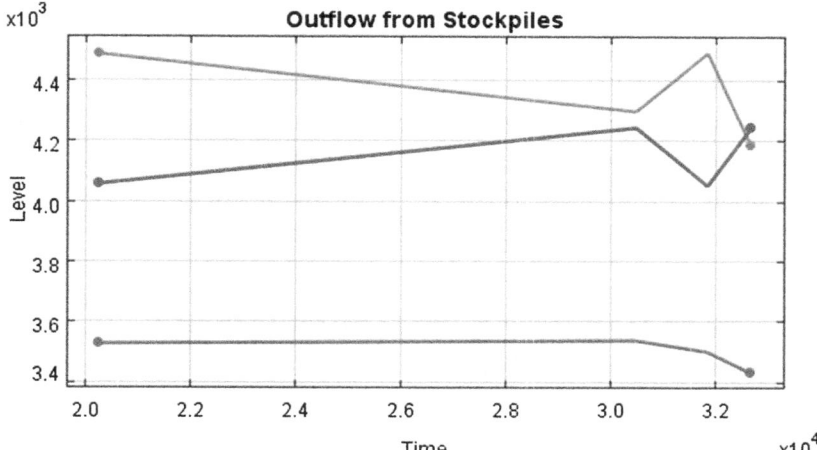

Fig. 4.28 Output of a simple simulation, outflow from the different stockpiles

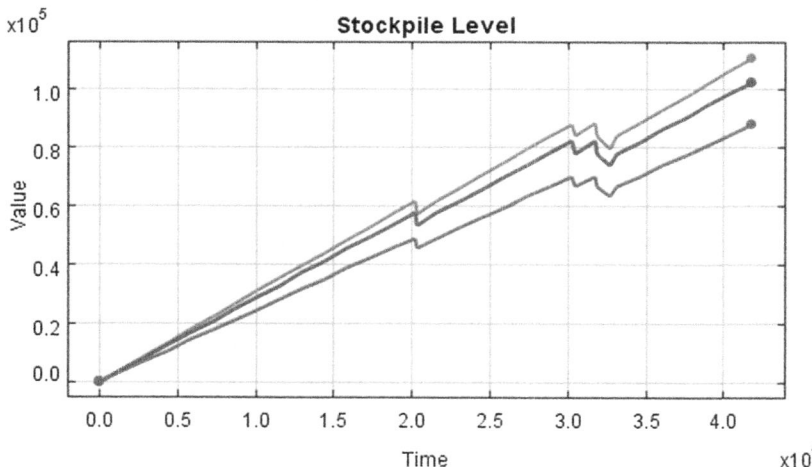

Fig. 4.29 Output of a simple simulation, evolution of the levels of the stockpiles over time

torical data to assess the consistency of the simulated values with reality. In order to provide an understanding of how the different components of the model interconnect, an overview will show only one possible connection path. This will allow the reader to visualise the concept behind the model. In order to simplify the presentation and construction of the model, and given the absence of relative pricing simulations in the DES model, it was decided to omit modelling of the outbound logistics.

4.9 Note Regarding Ptolemy II Images Used in the Text

The Ptolemy II images and screen-shots used in the chapter are subject to the following license:

> Copyright ©1990–2017 The Regents of the University of California. All rights reserved.
>
> Permission is hereby granted, without written agreement and without license or royalty fees, to use, copy, modify, and distribute this software and its documentation for any purpose, provided that the above copyright notice and the following two paragraphs appear in all copies of this software.
>
> IN NO EVENT SHALL THE UNIVERSITY OF CALIFORNIA BE LIABLE TO ANY PARTY FOR DIRECT, INDIRECT, SPECIAL, INCIDENTAL, OR CONSEQUENTIAL DAMAGES ARISING OUT OF THE USE OF THIS SOFTWARE AND ITS DOCUMENTATION, EVEN IF THE UNIVERSITY OF CALIFORNIA HAS BEEN ADVISED OF THE POSSIBILITY OF SUCH DAMAGE.
>
> THE UNIVERSITY OF CALIFORNIA SPECIFICALLY DISCLAIMS ANY WARRANTIES, INCLUDING, BUT NOT LIMITED TO, THE IMPLIED WARRANTIES OF MERCHANTABILITY AND FITNESS FOR A PARTICULAR PURPOSE. THE SOFTWARE PROVIDED HEREUNDER IS ON AN "AS IS" BASIS, AND THE UNIVERSITY OF CALIFORNIA HAS NO OBLIGATION TO PROVIDE MAINTENANCE, SUPPORT, UPDATES, ENHANCEMENTS, OR MODIFICATIONS.

Notes

1. Please recall that the rai is the unit of measurement used in Thailand; any other alternative measurement (hectare, acre, etc.).
2. Please note that the model needs to be run at least once for the `SetVariable` actor to store at least a value. If nothing occurs during the first run of the model, it must be run again to store a value. This is one of the reasons why a parameter with the same name as the variable is used in the composite model. It is a means of initialising a value and prevents the raising of an exception.
3. This composite model has been inspired by one of the examples found in Ptolemy II.
4. `capacity` is initialised to the `Capacity` name defined in the scope of the composite actor, and actually a named parameter for it.

5
Model Implementation and Validation

Abstract In this chapter the building blocks developed in Chap. 4 are used to model the Southern Thailand Rubber supply chain. After the model is built, it is validated to check the consistency of the observed movement and the material movement at the regional level. The chapter concludes by analysing the capability of the model by using two case studies.

Keywords Discrete event simulation · The thai rubber industry Ptolemy II

This chapter takes the building blocks from Chap. 4 and develops them into a model for the rubber industry supply chain in Thailand's southern region. After the model is built, it will be validated to check that material movements are consistent with observed movements at the regional level. Finally, two case studies will be analysed and the capabilities of the model will be illustrated.

5.1 Model Implementation

5.1.1 Data

In order to implement the model, we need to base it on existing data. In this case, the model is Southern Thailand's rubber supply chain. This area is composed of 14 provinces, which are listed in Table 5.1, and whose geographical placement can be observed in Fig. 5.1.

Recall that according to our DES model, every province is characterised by the following parameters:

- ` Productivity per rai`
- ` Number of farmers`
- ` LX Proportion`
- ` US Proportion`
- ` CL Proportion`

Table 5.1 Rubber production in each province in year 2015 (adjusted by authors, based on data from (TTRA 2016))

Province	Cultivation area (rai)	Total yield (kg)	Average yield[a] (kg/rai)	LX %	CL %	US %
Chumporn	503,594	132,445,222	263	30	30	40
Surat Thani	2,571,532	699,456,704	272	30	40	30
Nakhon Si Thammarat	1,666,416	426,602,496	256	40	30	30
Krabi	693,216	158,746,464	229	50	20	30
Trang	1,321,658	303,981,340	230	50	20	30
Phangnga	580,050	139,212,000	240	25	25	50
Phuket	77,869	17,676,263	227	50	25	25
Ranong	218,548	48,736,204	223	25	50	25
Songkhla	1,798,306	487,340,926	271	30	30	40
Satun	355,302	84,561,876	238	30	40	30
Phattalung	675,692	166,220,232	246	30	40	30
Pattani	292,795	70,856,390	242	50	25	25
Yala	1,148,968	276,901,288	241	50	25	25
Narathiwat	881,359	195,661,698	222	50	25	25

[a] http://www.oae.go.th/download/prcai/farmcrop/rubber.pdf

Fig. 5.1 Southern Provinces of Thailand (created by the authors)

The productivity per rai and production proportions for LX, US and CL can be obtained from Table 5.1. For the number of farmers, sample data for each province of farmers and their size (in rai) has been obtained. From there a distribution will be modelled for each province. From that model, the number of farmers that provide a representative production amount for a period of one month can be implemented.

5.1.2 Modelling the Number of Farmers

Initially, we attempted to adjust the distribution of the samples available, but due to the incompleteness of the data it was not possible to have a detailed description of both farmer numbers and size distribution. The process applied consisted of assuming a model (power law model) to try to fit the parameters for this distribution based on the sample available. The output of the process can be seen in Table 5.2.

The histogram seems to indicate long tails for the distributions, and within some areas there seems to be a power law behaviour as shown in Fig. 5.2. Unfortunately, the lower tail of the distribution is not well approximated by a power law. However, the tails that contribute the most

Table 5.2 Parameters for modelling farmer size distributions in each province

Province	$x_{\min}$ (rai)	$x_{\max}$ (rai)	α
Chumporn	1.1	381	4.2
Surat Thani	1.0	125	6.1
Nakhon Si Thammarat	0.8	120	3.5
Krabi	1.3	90	4.6
Trang	1.3	154	3.6
Phangnga	1.5	185	3.3
Phuket	1.4	348	3.0
Ranong	2.0	88	3.5
Songkhla	0.6	150	3.7
Satun	0.4	163	3.8
Pathalung	1.0	66	3.5
Pattani	0.8	126	3.5
Yala	2.0	134	3.4
Narathiwat	0.5	100	3.4

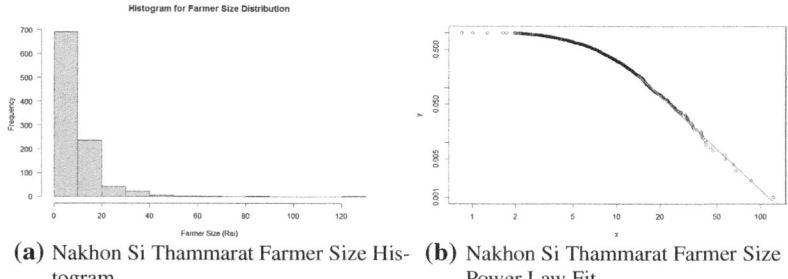

(a) Nakhon Si Thammarat Farmer Size Histogram (b) Nakhon Si Thammarat Farmer Size Power Law Fit

Fig. 5.2 Nakhon Si Thammarat Province

(right tails of distribution) are more precisely modelled. It will be assumed that the distributions for the farmer sizes follow a power law. For the histograms and power law fit for all provinces, see Appendix A. Note that the information available is incomplete, so we have taken a sample (at random) of what was available and modelled the power law distribution based on this. The next step in the process is to calculate how many farmers are needed (on average) so that the production of each province can be approximately matched by the production of the model. For this purpose, a small simulation model was set up to determine the approximate number of farmers. The idea of the model is to propose different numbers of farmers, generate their sizes according to the corresponding power law, examine the production outcome and attempt to match that of the province (taking into account productivity per rai).

The model that was built essentially takes the elements already used in the model described in Chap. 4 but with some minor modifications. The model can be seen in Figs. 5.3 and 5.4. The modifications added were that the clock is fired every unit of time and the DEDirector is left to run 100 times. For this reason an Average actor was added to calculate the average yield due to the 100 realisations generated.

This process is manual and involves some trial and error. It has also been observed that the time required to run the simulation can take several minutes if the number of farmers is too large. Consequently, it was decided that a factor would be applied to the models to avoid excessive running times. A factor of 1000 in the cultivation area was applied to every province

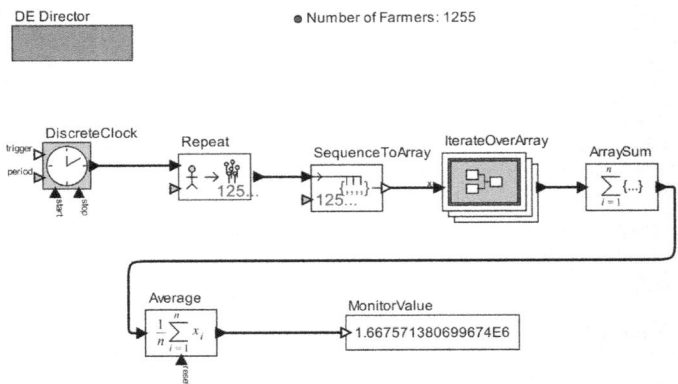

Fig. 5.3 Model used for calibration

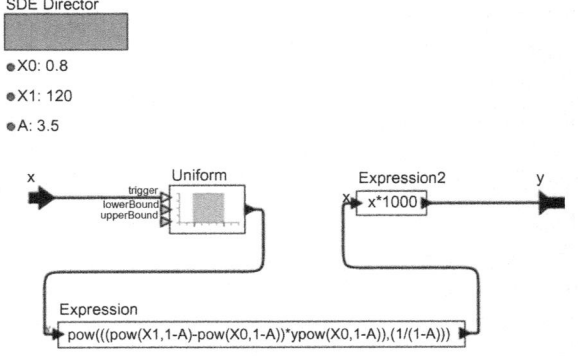

Fig. 5.4 Detail of the change in productivity to reduce the execution time inside the `IterateOverArray` actor

model in order to reduce the running time by orders of magnitude. An average of 100 realisations was taken in order to determine the approximate number of farmers. By doing this, some confidence numbers were possible and the impact of a particular realisation was avoided (see Table 5.3). Finally, for the application of the model, it needs to be noted that the productivity per rai available in Table 5.1 is the average yield per year; consequently, this number is divided by 365 to obtain average productivity per day.

5.1 Model Implementation

Table 5.3 Approximate number of farmers in each province

Province	Approximate number of farmers (Unit: 1000)	Rubber Plantation area (rai)
Chumporn	316	505,871.33
Surat Thani	2070	2,575,704.43
Nakhon Si Thammarat	1255	1,666,372.37
Krabi	387	695,475.42
Trang	627	1,321,292.24
Phangnga	222	587,914.77
Phuket	28	79,558.54
Ranong	66	219,158.94
Songkhla	1887	1,799,192.03
Satun	573	356,324.35
Pathalung	408	678,628.48
Pattani	220	292,997.46
Yala	337	1,148,970.26
Narathiwat	1033	881,202.08

A validation example was run. For this example a model was built (see Fig. 5.5). This example was run for a full year, which corresponds to 525600 min of simulation time.

The previous model was built by adding a `DEDirector`, which was set to 525600 as the running time. Composite actors for the provinces (the same already modelled in the previous chapter) were responsible for inputting the data for each one. An `Accumulator` actor that calculates the sum of everything was input to it (in this case all the rubber produced). Finally, a `MonitorValue` actor showed the value that was produced as an output of the `Accumulators`. Table 5.4 contains a comparison between the data and the output of the model.

It is important to note that the clock used on the previous model was a `DiscreteClock`, which sends regular tokens at specified intervals. This actor, unfortunately, fires on initialisation of the model, hence the need to modify the `offsets` parameter (initialised at `0.0`) to `1440.0` (remember that 1440 is the number of minutes every day). By doing this, the initial signal that would otherwise fire at `0.0` now fires at `1440.0`. Note that the parameter cannot be set to a number higher than `period`.

104 5 Model Implementation and Validation

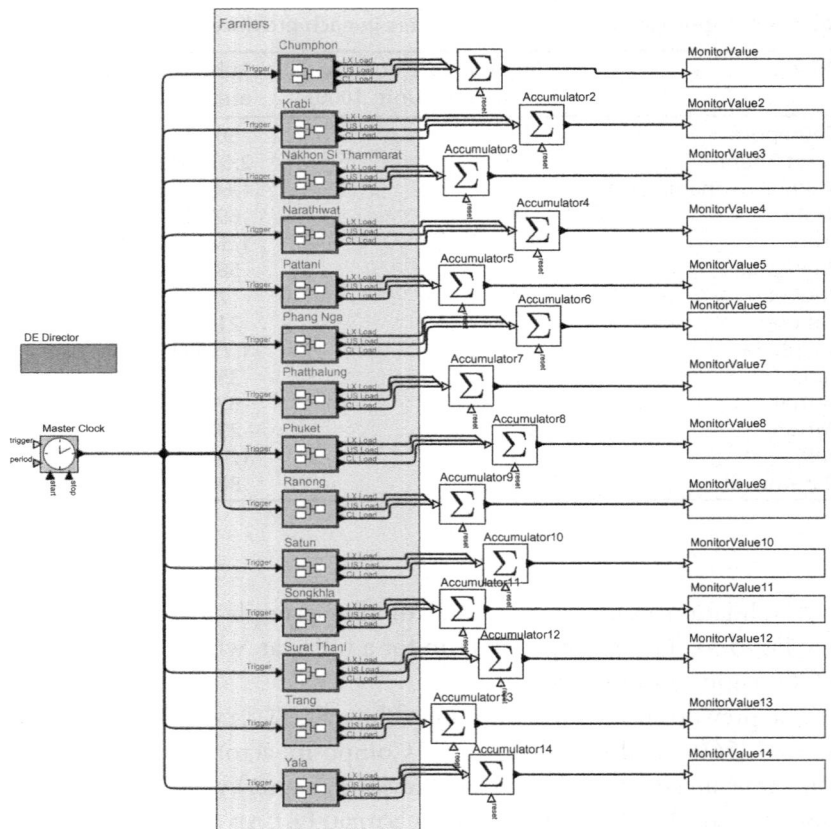

Fig. 5.5 Model implemented for testing total product generation in each province for 1 year

It can be seen that the values produced by the simulation model are very close to the production figures for 2015. If the model is run with the current time setup, it takes a considerable amount of time (it will depend on the machine used). In future runs it is recommended that rather than using a whole year for the simulation exercise, a month is used instead. A month can be approximated by $30 \times 24 \times 60 = 43200$ min. If the model is run with that time, then the execution time of the model should be much more manageable than for the full year.[1]

Table 5.4 A comparison of data values and simulated values for farmer production in each Province in 2015

Province	Production (rai)	Model production (rai)
Chumporn	132,445,222	133,005,361.41
Surat Thani	699,456,704	700,319,018.11
Nakhon Si Thammarat	426,602,496	426,765,388.81
Krabi	158,746,464	159,412,846.07
Trang	303,981,340	303,979,904.70
Phangnga	139,212,000	141,723,176.92
Phuket	17,676,263	17,693,989.82
Ranong	48,736,204	48,664,358.24
Songkhla	487,340,926	486,648,051.80
Satun	84,561,876	84,892,838.78
Phattalung	166,220,232	166,884,524.48
Pattani	70,856,390	70,874,847.28
Yala	276,901,288	277,679,373.65
Narathiwat	195,661,698	196,863,176.66

5.1.3 Modelling the Distribution Centres

Distribution centres as such should not be difficult to model. In the previous chapter there was a preliminary model for the distribution centre, and it will be assumed (for the sake of simplicity) that one representative distribution centre is available per province with a capacity that represents that of the aggregated distribution centres of the province plus some additional allowance. Figure 5.6 provides an overview of the model after adding the distribution centres layer to it.

Table 5.5, shows the storage requirements per province obtained from the production of 2015 (see Table 5.1. The value obtained from Table 5.1 has been divided by 12 (to represent months) and an allowance of 5% has been applied.

At first glance it seems that having the distribution centres in the model is superfluous, as they do not impose constraints on the operation of the supply chain. Nevertheless, in future modelling, cost and emission considerations will very likely make this component of the supply chain important. For the moment, the focus must stay on building the model. After this stage is successfully completed, changes may be effected in order to

106 5 Model Implementation and Validation

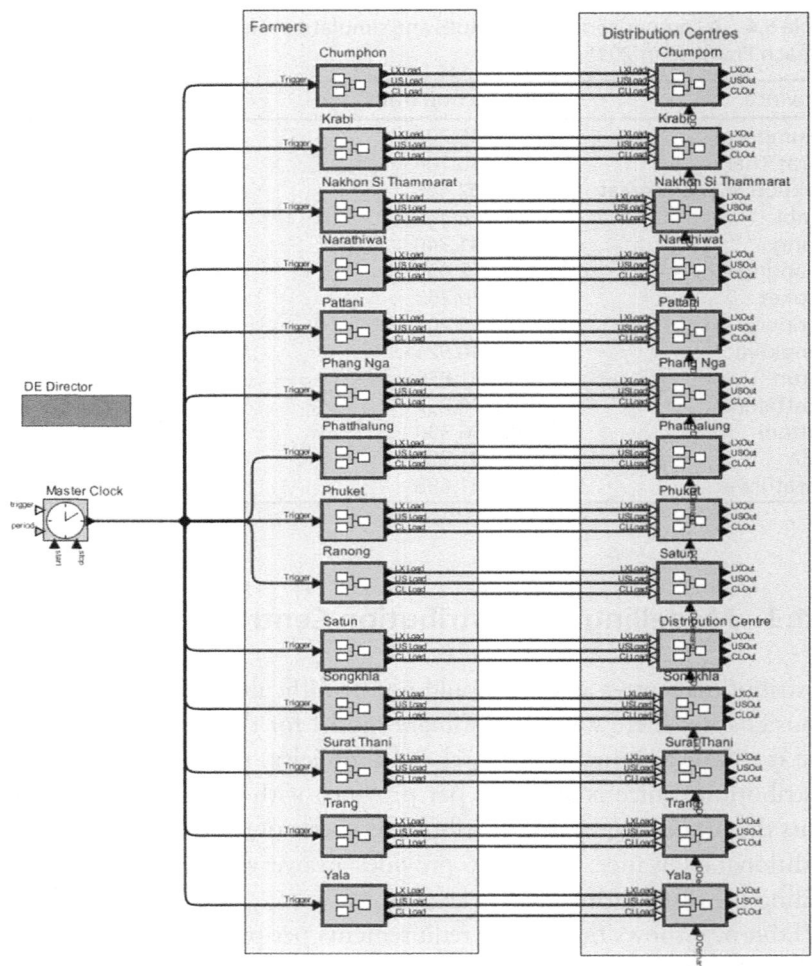

Fig. 5.6 Model implemented including up to distribution centres

study the possibility of any problems imposed on the operation of the supply chain by the distribution centres.

Table 5.5 Distribution centre capacity in each Province

Province	Monthly storage capacity (kg)	5% Allowance (kg)	Total (kg)
Chumporn	11,037,083.33	551,854.17	11,588,937.50
Surat Thani	58,288,083.33	2,914,404.17	61,202,487.50
Nakhon Si Thammarat	35,550,166.67	1,777,508.33	37,327,675.00
Krabi	13,228,833.33	661,441.67	13,890,275.00
Trang	25,331,750.00	1,266,587.50	26,598,337.50
Phangnga	11,601,000.00	580,050.00	12,181,050.00
Phuket	1,473,000.00	73,650.00	1,546,650.00
Ranong	4,061,333.33	203,066.67	4,264,400.00
Songkhla	40,611,750.00	2,030,587.50	42,642,337.50
Satun	7,046,833.33	352,341.67	7,399,175.00
Phattalung	13,851,666.67	692,583.33	14,544,250.00
Pattani	5,904,666.67	295,233.33	6,199,900.00
Yala	23,075,083.33	1,153,754.17	24,228,837.50
Narathiwat	16,305,166.67	815,258.33	17,120,425.00

5.1.4 Modelling the Factories

For the factories, the data available are summarised in Table 5.6. At first, the total number of factories is considered, but a more detailed consideration is given at a later stage. Some provinces do not have any factories, and the number of factories differs between provinces. At the time of writing, there is no information available regarding the processing capacity of each factory. Therefore, at this point in the modelling it is assumed that capacity is not a factor and that factories can process any amount desired. It is acknowledged that this assumption is a strong one; however, it is functional for the purposes of simplicity for the construction of the first model of the supply chain and for using this for post-validation.

It can be noted that one of the concepts used in implementing the modelling of a factory composite actor, as described in the previous chapter, is relevant. Factories "pull" material from distribution centres, i.e. they produce tokens that are passed backwards in the chain.

For the purposes of keeping the simulation simple, one factory actor will be used for each province and the demand will be applied to all these

Table 5.6 Number of factories in each province

Province	Number of factories			
	LX	US	CL	Total
Chumporn	3	5	2	10
Surat Thani	12	9	7	28
Nakhon Si Thammarat	7	21	13	41
Krabi	3	4	1	8
Trang	12	6	6	24
Phangnga	1	1	1	3
Phuket	1	0	2	3
Ranong	0	0	0	0
Songkhla	25	27	29	81
Satun	0	0	2	2
Patthalung	2	6	3	11
Pattani	1	3	5	9
Yala	5	4	3	12
Narathiwat	0	4	1	5

actors. However, the requirement will be then proportional to the number of factories that each province has.

Changes Required in the Factory Composite Actor

The initial factory actor was created with one representative factory actor per province. However, as the data show, there is a province that does not have a factory (Ranong). Furthermore, there is a possibility that some factories could obtain primary rubber products from distribution centres located outside the province but still within a reasonable distance of a particular factory. For these reasons, the factory actor needs to be changed so that it possesses a mechanism for assigning the factor demand to the different provinces.

To achieve this, one option is to have a vector of proportions for each province and then choose a sensitive value for those weights trying to balance the supply chain. It has already been mentioned that the underlying problem at every period of time is the economic quantity optimisation for each factory. However, without information on relative prices and

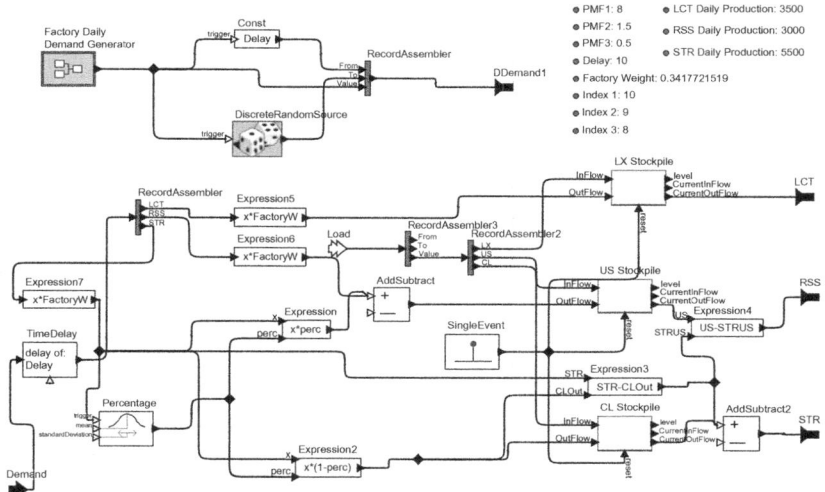

Fig. 5.7 Modified version of the factory composite actor that allows distribution of demand between provinces

transportation costs it is impossible to solve that problem, hence the need for simplification. One alternative to modelling where the factory obtains its material is to specify a discrete probability distribution with weights. For the sake of simplicity, the modified factory composite actor model will allow for a maximum of three output ports that will be assigned at random, following the discrete probability function. Figure 5.7 shows the modified model.

The new model incorporates a `DiscreteRandomSource` actor. This actor is of an advanced level, and it is suggested that the reader consult the documentation that comes with it, along with the examples. Essentially, the `DiscreteRandomSource` actor creates values based on a probability mass function. An example is provided in Fig. 5.8.

The parameters `PMF1`, `PMF2` and `PMF3` are set using parameters so that the user of the model can set them at the factory composite model level. Note that within the `DiscreteRandom` actor, the probability

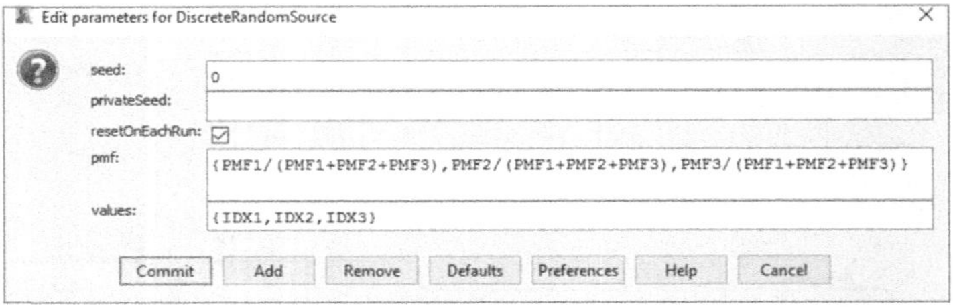

Fig. 5.8 Modified version of the factory composite actor that allows distribution of demand between provinces

masses are calculated based on those parameters using a formula of the type

$$f_i = \frac{PMFi}{PMF1 + PMF2 + PMF3}.$$

With this choice, it is automatic that the sum of the probability masses adds up to 1 (a requirement of the `DiscreteRandom` actor); otherwise, the software would raise an error.

Furthermore, to avoid clashes of tokens coming from different factories, a new parameter, `Delay`, can be introduced that creates a delay in the generation of the tokens. Note that this `Delay` parameter is also used to identify the factories and is a value in the set $\{0, \ldots, 13\}$. This is because later, the `Switch` actor needs to decide the output based on a number in the set $\{0, \ldots, n-1\}$, with n being the number of choices for the `Switch`.

Finally, an additional parameter called `Factory Weight` is added to the composite actor to take into account the demand distribution between factories. The factory weight is defined as the number of factories in the province divided by the total number of factories in the country.

Addition of a Dispatcher Actor

The critical change required for the final implementation of the model is the addition of a `Dispatcher` composite actor. The logic behind this actor is that of directing the token from a factory to a province

5.1 Model Implementation

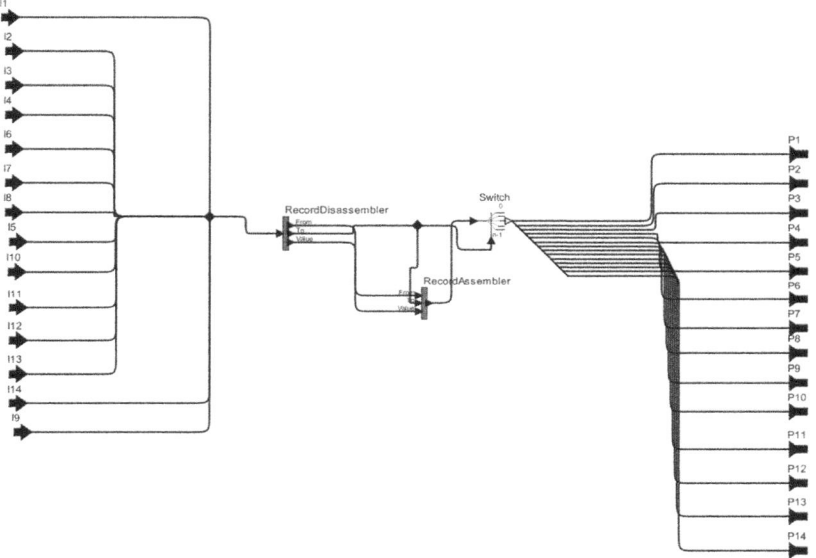

Fig. 5.9 Implementation of the dispatcher actor

based on the From and To values that are encoded in the token (using the RecordAssembler). In addition, a Switch actor is used. The Switch actor will direct a token to an output port (numbered from 0 to $n-1$) based on the index passed to it as a parameter. The implementation of this actor is presented in Fig. 5.9.

Changes Required in the Distribution Centre Composite Actor

As the origin and destination, together with the value, are encoded using an actor named RecordAssembler, it is necessary to read this input, process it and re-direct it back to the sender. In order to achieve this, the distribution centre composite actor needs to be changed.

The main change introduced in the actor relates to the encoding used by the token that needs to be disassembled, processed and assembled again,

112   5 Model Implementation and Validation

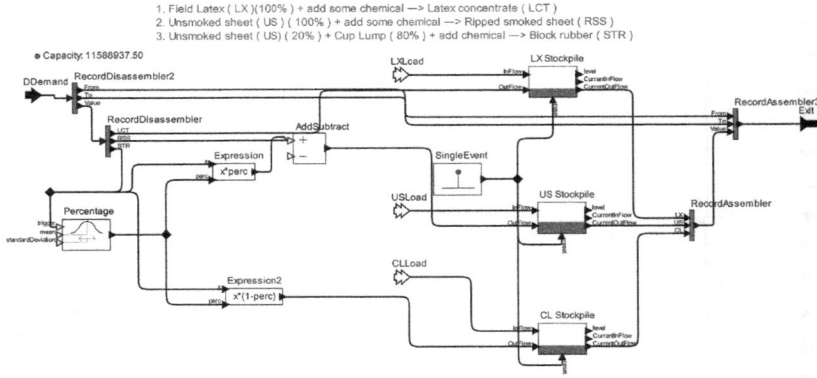

Fig. 5.10 Modified version of the distribution centre composite actor that allows the directing of tokens back to factories

taking care to invert the *origin-destination* pair provided with the values. By doing this, it is possible to direct the token back to its originator. Apart from this, the change is quite straightforward and has minimal impact on the original actor's logic. This modification is presented in Fig. 5.10.

5.1.5 Modelling the Demand

For the implementation of the model there is no change in the current model for the demand. However, other possibilities could be used in the future. For example the demand for secondary rubber products could be modelled following distributions other than normal (e.g. the power law). Unfortunately, at the time of this writing, the demand process is not properly characterised and it is difficult to guess what form it takes. For the purposes of the model, the existence of a demand composite actor that can later be replaced by other more complex actors or a more accurate implementation is good enough.

5.1.6 General Overview of the Model

Figure 5.11 contains an overview of the whole model as implemented.

5.2 Model Validation

For the purposes of validating the model, it is necessary to "extract" information after the model has been run. The model currently exposes the `LCT`, `RSS` and `STR` coming out of the factory models. However, they are outputs for each factory, hence the need to aggregate them. An `Accumulator` actor can be used for this purpose as depicted in Fig. 5.12.

When the model is run, the graphical output contained in Fig. 5.13 is produced.

As can be seen, after the demand is produced, the different outputs of secondary rubber products from the provinces arrive. The `Timed Plotter` appears to show everything occurring at the same time. However, when performing a zoom (see Fig. 5.13d), it can be observed that the outputs from the provinces arrive at regular intervals, the exception being province number nine which corresponds to Ranong, the province that does not have factories.

As mentioned, in this model a `Delay` parameter was used to identify provinces but also to space out the tokens to avoid their overlapping in terms of time-stamps. The output shows that under an external demand the model produces outputs coming from all provinces to satisfy that demand. This confirms the correctness of the model from an implementation point of view. For future case studies, more realistic demand data (contained in Table 5.7) will be used. However, there is no formal source for it, so it needs to be considered with care. Note that the daily demand for factories has been approximated as the daily demand with an additional 10% applied.

The daily demand for the factories is summarised in Table 5.8. This demand was built taking into account the daily average external demand and then using the factory multiplier.

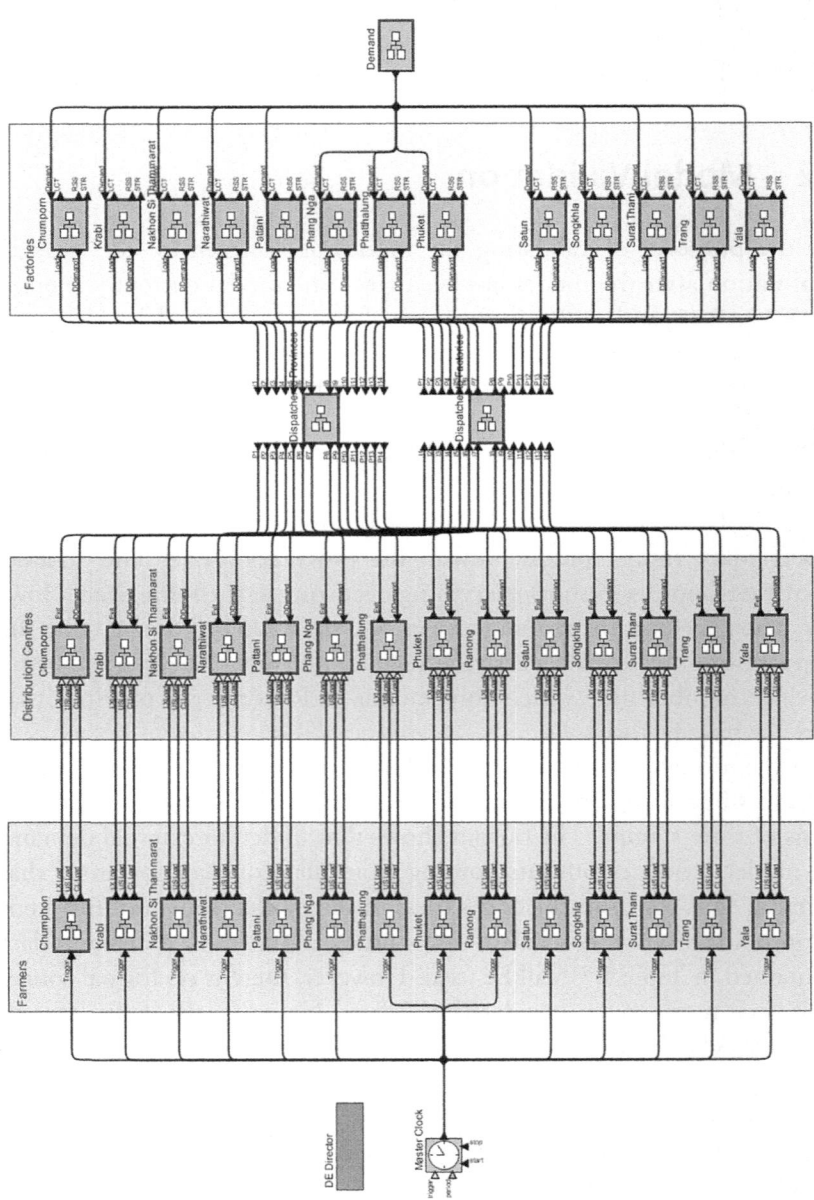

Fig. 5.11 An overview of the complete model

5.2 Model Validation 115

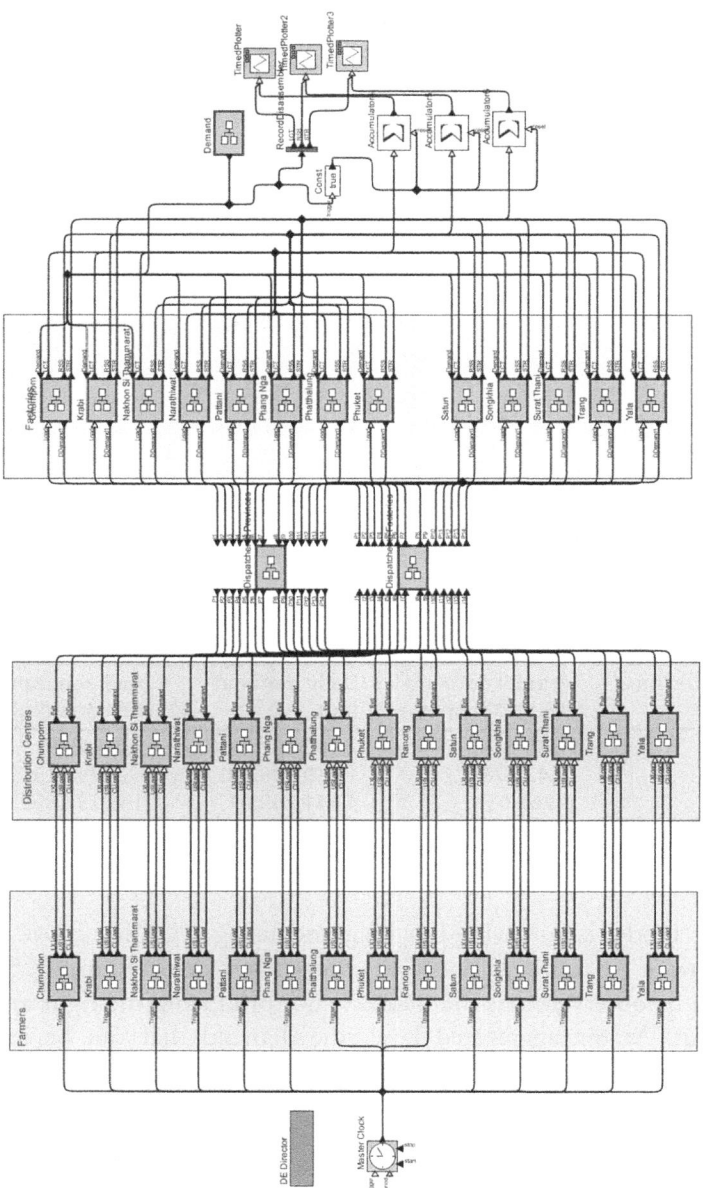

Fig. 5.12 An overview of the complete model with the extra actors to capture information

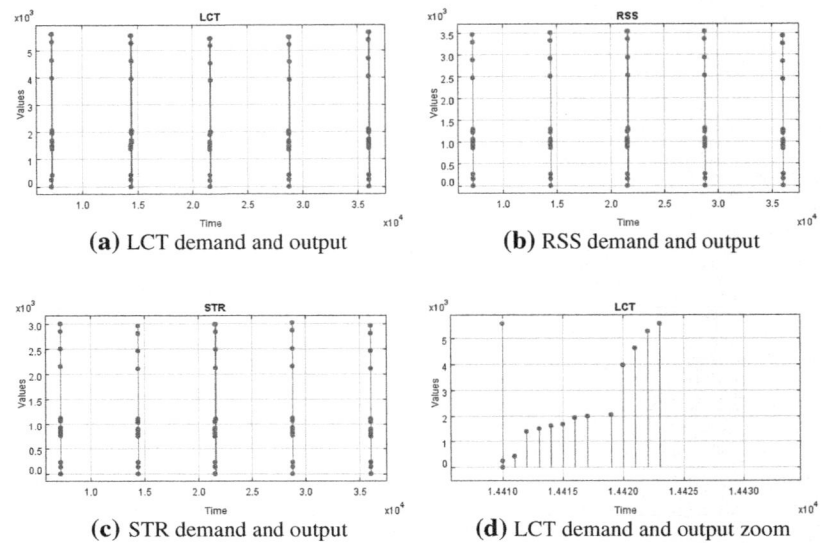

Fig. 5.13 Graphical output of a run of the model

Table 5.7 Rubber demand data (adjusted by authors, based on data from (TTRA 2016))

Rubber Product	Annual demand (Metric Tonnes)	Daily demand (kg)	Average monthly demand (kg)
LCT	730,364	2,000,997.26	60,029,917.80
RSS	642,378	1,759,939.73	52,798,191.90
STR	1,767,061	4,841,263.01	145,237,890.30

If the model is run with this data and using a DiscreteClock rather than a PoissonClock, the output is as depicted in Fig. 5.14.

It can be observed that the model is not producing the right amount of output. At the aggregated level, the demand that can be implied from the secondary rubber products is in total 3,139,803,000 kg, while the aggregated production is 3,208,397,000 kg. For secondary rubber product demand as presented previously, the primary rubber products required should be 730,364,000 kg of LX, 995,790,200 kg of US and 1,413,648,800 kg of CL. Based on the current production weights assigned

Table 5.8 Factory demand data (adjusted by authors, based on data from (TTRA 2016))

Province	LCT (kg)	RSS (kg)	STR (kg)
Chumporn	84,430.264	74,259.06	204,272.701
Suratthani	236,404.74	207,925.369	571,963.562
Nakhon Si Thammarat	346,164.083	304,462.147	837,518.073
Krabi	67,544.211	59,407.248	163,418.161
Trang	202,632.634	178,221.745	490,254.482
Phangnga	25,329.079	22,277.718	61,281.81
Phuket	25,329.079	22,277.718	61,281.81
Ranong	0	0	0
Songkhla	683,885.139	601,498.389	1,654,608.877
Satun	16,886.053	14,851.812	40,854.54
Patthalung	92,873.291	81,684.966	224,699.971
Pattani	75,987.238	66,833.154	183,845.431
Yala	101,316.317	89,110.872	245,127.241
Narathiwat	42,215.132	37,129.53	102,136.35

to each province, the total of each primary rubber product will be 1,200,546,300 kg of LX, 1,021,848,550 kg of US and 986,002,150 kg of CL. According to this, there should only be a shortage of STR, as the other two (LCT and RSS) should be greater than that supplied. Some of the discrepancies between the simulated output and the desired output can be explained by the use of production multipliers for farmers at the province level, but there is still more missing information that would explain all the differences. The other potential source of problems relates to the eventual discrepancy that exists between the sourcing of primary rubber products for factories and the size of the demand for the factories in the province. In short, the demand flows are not balanced and it is not feasible to supply the desired output with the current configuration.

In order to alleviate these problems, the following parameters listed in Table 5.9 are suggested. It needs to be noted at this point that the nature of the demand and the relative importance of the provinces in terms of factories make the problem of finding a feasible assignation a difficult one.

Table 5.9 Demand from factories to provinces (adjusted by authors, data from (TTRA 2016))

	Chumporn (%)	Suratthani (%)	Nakhon Si Thammarat (%)	Krabi (%)	Trang (%)	Phangnga (%)	Phuket (%)	Ranong (%)	Songkhla (%)	Satun (%)	Phattalung (%)	Pattani (%)	Yala (%)	Narathiwat (%)
Chumporn	74.4	0	0	0	0	0	0	0	3	0	0	0	0	0
Suratthani	0	82.7	25.5	0	0	0	0	0	22.5	0	0	0	0	0
Nakhon Si Thammarat	0	11.2	70.1	0	0	0	0	0	0	0	0	0	0	0
Krabi	0	0	4.4	1	5	0	0	0	0.7	0	0	0	0	0
Trang	0	0	0	0	85.8	0	0	0	2.8	0	0	0	0	0
Phangnga	0	1.8	0	0	0	1	66.3	0	0.6	0	0	46.6	0	0
Phuket	0	0.5	0	0	0.6	0	33.7	0	0	0	0	0	0	0
Ranong	25.6	3.8	0	0	0	0	0	0	0	0	0	0	0	0
Songkhla	0	0	0	0	0	0	0	0	44.7	0	0	0	0	0
Satun	0	0	0	0	0	0	0	0	5	1	0	0	0	0
Phattalung	0	0	0	0	0	0	0	0	1.5	0	1	0	0	0
Pattani	0	0	0	0	0	0	0	0	0.7	0	0	53.4	0	0
Yala	0	0	0	0	8.7	0	0	0	7.6	0	0	0	1	0
Narathiwat	0	0	0	0	0	0	0	0	10.8	0	0	0	0	1

5.2 Model Validation

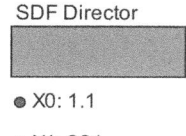

- X0: 1.1
- X1: 381
- A: 4.2

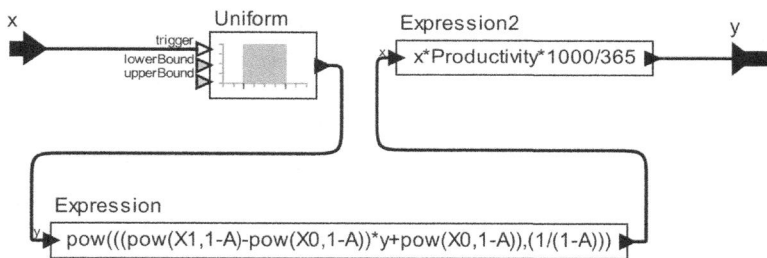

Fig. 5.14 Graphical output of a run of the model with new data

An optimisation model was attempted which had the form:

$$\min \sum_{i=1}^{N} \sum_{j=1}^{N} C_{i,j} P_{i,j}$$

s.t.

$$\sum_{j=1}^{N} P_{i,j} = 1 \quad \forall i \in \{1, \ldots, N\} \tag{5.1}$$

$$\sum_{i=1}^{N} Q_i \cdot P_{i,j} \geq D_j \quad \forall j \in \{1, \ldots, N\} \tag{5.2}$$

$$P_{i,j} \geq 0 \quad \forall i, j \in \{1, \ldots, N\} \times \{1, \ldots, N\} \tag{5.3}$$

$$P_{i,j} \leq 1 \quad \forall i, j \in \{1, \ldots, N\} \times \{1, \ldots, N\} \tag{5.4}$$

with:

- N the number of provinces under consideration (for the particular example studied so far $N = 14$)
- $C_{i,j}$ the average transportation cost between distribution centre located at province i and factory j
- $P_{i,j}$ the proportion of production from province i that is taken by factories in province j
- Q_i the total quantity offered by province i
- D_j the demand of factories in province j (Ranong as a province does not have factories so in this case D_j is zero)

Constraint 5.1 ensures that what is distributed from every province is equal to 100%. This condition can actually be relaxed a bit to

$$\sum_{j=1}^{N} P_{i,j} \leq 1 \ \forall i \in \{1, \ldots, N\}.$$

Constraint 5.2 is asking the model to satisfy the factories' demand in each province (recall that currently what is being done is to take an external demand and apply it proportionately to the number of factories each province has). Constraints 5.3 and 5.4 simply indicate the nature of the variables.

The model was attempted, but due to the nature of the constraints imposed a priori, such as factories of a province not being able to source from more than three provinces, the model was not able to show any feasible solution. By using a trial-and-error process, it was found that one particular province (Songkhla) required much more than just three provinces to feed its factories (in fact, it is the province with the largest factory capacity in Southern Thailand). Another important thing to observe at this point is that, depending on how we look at it, the number $P_{i,j}$ can mean two different things. The quantity

$$\sum_{j=1}^{N} P_{i,j}$$

is always equal to one, and as mentioned before this means that distribution centres in a province will not send more material than they possess. However, some provinces like Songkhla receive material from several other provinces, which causes the sum

$$\sum_{i=1}^{N} P_{i,j}$$

to be greater than one in some cases. This should not be a problem. However, the way the split mechanism has been implemented, it takes what factories are demanding and distributes this to distribution centres in different provinces. Consequently, the numbers needed for the model are the normalised weights (which are the ones reported in Table 5.9), i.e.

$$\tilde{P}_{i,j} = \frac{P_{i,j}}{\sum_{j=1}^{N} P_{i,j}}.$$

Because of this problem, a small modification is needed in the `DiscreteRandomSource` actor for each factory composite model. Previously it was modelled by considering only three source provinces, but as mentioned earlier, this did not produce feasible solutions, hence the need to extend the number of provinces considered. A list considering all possible provinces is then created, which, for example in the case of Songkhla, is for the `pmf`:

```
{0.03,0.225,0,0.007,0.028,0.006,0,0,0.447,
0.05,0.015,0.008,0.076,0.108}
```

and for the `values` is

```
{0,1,2,3,4,5,6,7,8,9,10,11,12,13}
```

If the model is run with this set of parameters, the output of Fig. 5.15 is obtained. It can be noted that there is still a difference in what is expected. This can only be due to the production percentages used. If production percentages are adjusted to those contained in Table 5.10 the values

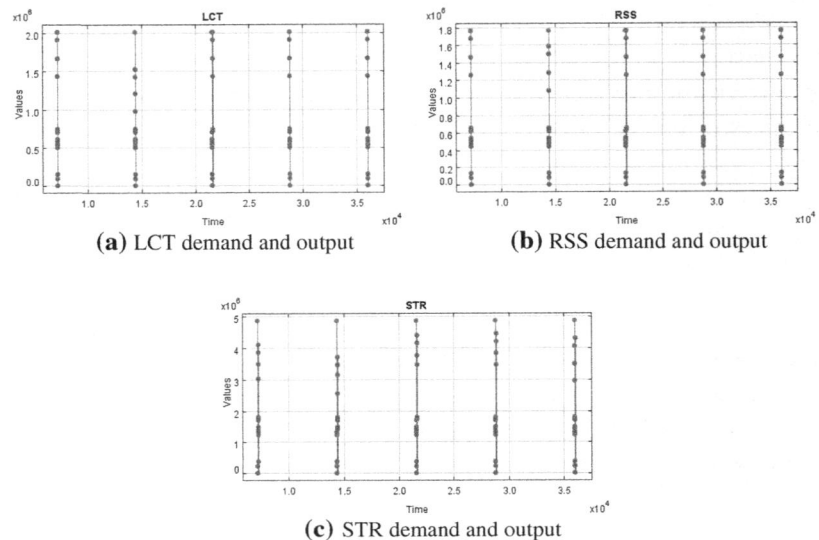

Fig. 5.15 Graphical output of a run of the model with new data

Table 5.10 New production percentages for each province

Province	LX (%)	CL (%)	US (%)
Chumporn	25	30	45
Suratthani	20	40	40
Nakhon Si Thammarat	25	30	45
Krabi	40	20	40
Trang	30	30	40
Phangnga	25	25	50
Phuket	40	25	35
Ranong	25	50	25
Songkhla	20	30	50
Satun	25	40	35
Phattalung	25	40	35
Pattani	25	25	50
Yala	20	25	55
Narathiwat	20	25	55

obtained get much closer to demand satisfaction, however it is not possible to obtain a model that achieves a real replication of the demand.

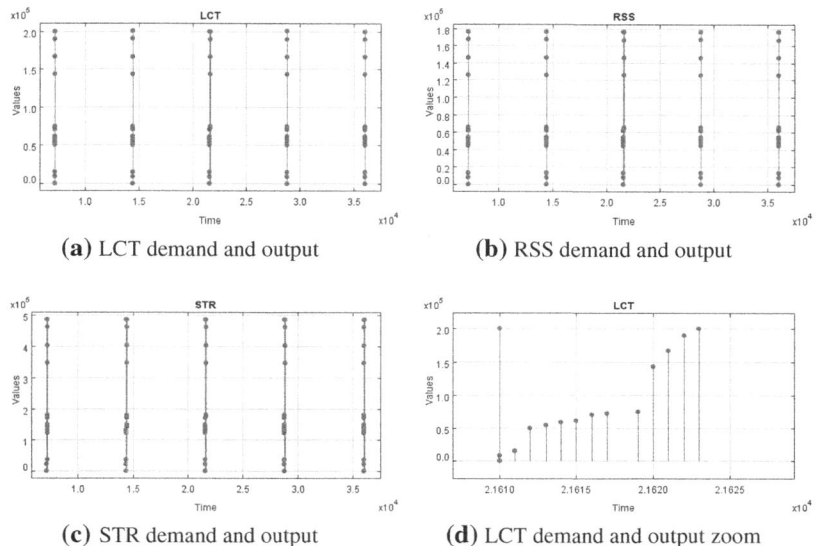

Fig. 5.16 Graphical output of a run of the model with demand data scaled down by one order of magnitude

As it is not possible to get demand satisfaction and the reasons for this problem are not yet clear, there are still several possible reasons for the lack of balance in the supply chain network. Unfortunately, without quality data, demand has to be lowered down by one order of magnitude (i.e. divided by 10). The model was run and the results obtained are summarised in Fig. 5.16.

5.3 First Case Study: Transportation Costs in the Supply Chain

In this case study the base model developed in the previous section will be used to calculate transportation costs in the supply chain. The model needs to be adjusted to calculate the transportation costs, but the modification is simple. The average transportation cost within a given province is assigned to the distribution centres within the same province. The cost between provinces (which affects the transportation cost to factories in the same and other provinces) is given in Table 5.11. Costs have been calculated by the authors as: Costs of transport products from each province = cost per kg-km x distance travelled in each province, based on data from (Shipping 2012).

In order to calculate the transportation costs, the model needs to account for the materials being moved to and from the provinces. The current token is a coding of origin, destination and value (coded and encoded using the previously described `RecordAssembler` and `RecordDissasembler` actors). This requires accessing a cost table and looking for an origin destination pair on it in order to apply the unit cost to the quantity transported. The identification of the cost can be performed using an advanced actor available in Ptolemy II called a `Python` actor, which allows the modeller to write simple actors using the Python programming language. `Python` itself is an advanced actor, and with documentation currently sparse, it is advisable to use other actors (if available) before resorting to `Python`. `JavaScript` is a similar actor that may be used for the same purposes using the JavaScript programming language.

The Python actor is added to the `Dispatcher` composite actor and configured with four ports: three inputs and one output. The three input ports are `To`, `From` and `Value`, which correspond to the respective values encoded in the token being dispatched. Note that the value part of the token corresponds to the quantities of LX, US and CL being sent from distribution centres to factories in order to apply a transportation cost to these aggregated quantities (in more advanced models it could be possible to apply differentiated transportation costs). For this reason, the model needs to be modified as shown in Fig. 5.17. The reader needs to be aware that errors in the scripts are difficult to debug, and therefore it is strongly suggested that a working version of the code be developed in an external editor before incorporating it into the `Python` actor; otherwise the development experience may be frustrating.

5.3 First Case Study: Transportation Costs in the Supply Chain

Table 5.11 Transportation costs between and within provinces in baht per kilogram (adjusted by authors, based on data from (Shipping 2012))

	Chumporn	Suratthani	Nakhon Thammarat	Si Krabi	Trang	Phangnga	Phuket	Ranong	Songkhla	Satun	Phattalung	Pattani	Yala	Narath-iwat
Chumporn	2.97	10.098	16.236	17.952	21.846	15.642	19.998	5.478	26.466	29.238	22.176	30.624	32.868	35.574
Suratthani	10.098	2.97	6.93	8.316	11.814	7.458	11.286	7.854	17.094	19.272	12.474	21.78	23.76	26.928
Nakhon	16.236	6.93	2.97	8.118	6.996	10.428	12.078	14.85	10.164	13.332	6.072	14.784	16.83	19.998
Krabi	17.952	8.316	8.118	2.97	6.27	4.026	3.96	14.124	13.794	13.53	9.108	19.206	20.526	24.354
Trang	21.846	11.814	6.996	6.27	2.97	10.23	9.042	19.074	7.722	7.656	3.564	13.134	14.322	18.15
Phangnga	15.642	7.458	10.428	4.026	10.23	2.97	4.422	11.022	17.688	17.556	12.87	23.034	24.486	28.248
Phuket	19.998	11.286	12.078	3.96	9.042	4.422	2.97	15.378	16.764	15.312	12.408	22.176	23.166	27.126
Ranong	5.478	7.854	14.85	14.124	19.074	11.022	15.378	2.97	24.948	26.664	20.196	29.634	31.68	34.848
Songkhla	26.466	17.094	10.164	13.794	7.722	17.688	16.764	24.948	2.97	5.808	4.95	5.412	6.93	10.56
Satun	29.238	19.272	13.332	13.53	7.656	17.556	15.312	26.664	5.808	2.97	7.326	9.042	8.91	12.87
Pattalung	22.176	12.474	6.072	9.108	3.564	12.87	12.408	20.196	4.95	7.326	2.97	10.164	11.748	15.378
Pattani	30.624	21.78	14.784	19.206	13.134	23.034	22.176	29.634	5.412	9.042	10.164	2.97	2.508	15.378
Yala	32.868	23.76	16.83	20.526	14.322	24.486	23.166	31.68	6.93	8.91	11.748	2.508	2.97	4.026
Narathiwat	35.574	26.928	19.998	24.354	18.15	28.248	27.126	34.848	10.56	12.87	15.378	5.28	4.026	2.97

126 5 Model Implementation and Validation

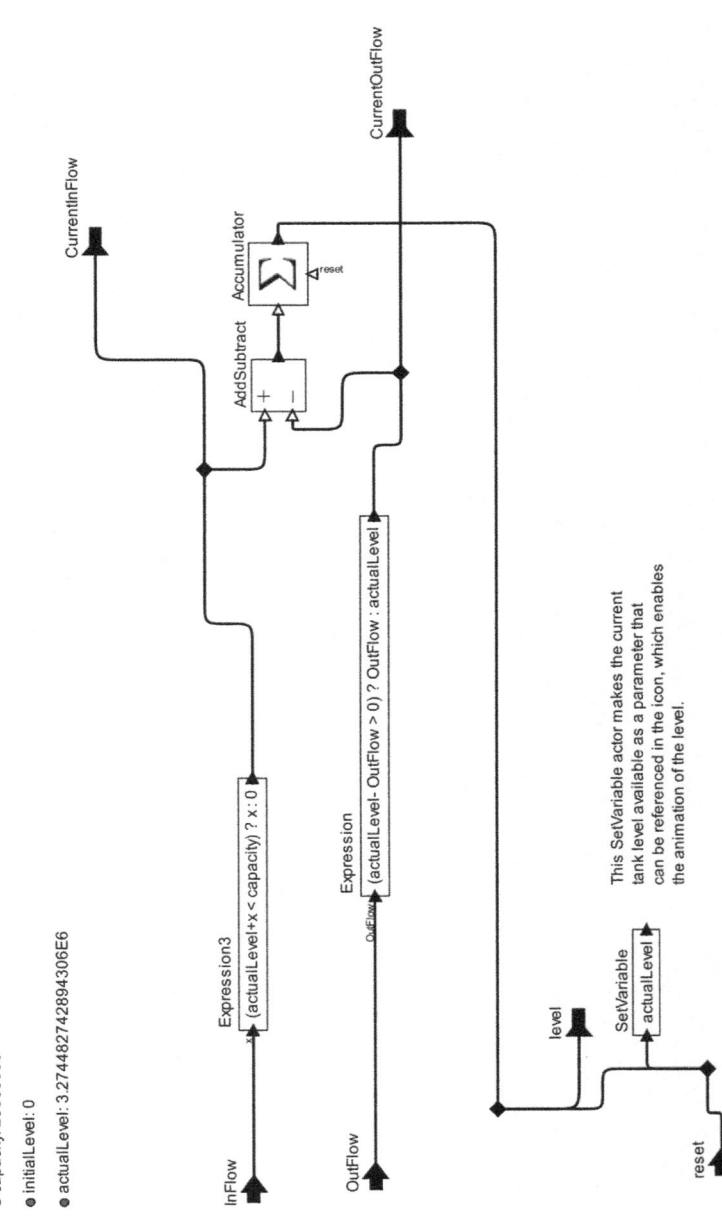

Fig. 5.17 Changes made to the Dispatcher actor to account for transportation cost from distribution centres to factories

5.3 First Case Study: Transportation Costs in the Supply Chain

The code contained in the Python actor is the following:

```
import ptolemy.data

class Main:

  def __init__(self):
    self.costToFromKg = [
      [2.97,10.098,16.236,17.952,21.846,15.642,19.998,5.478,
      26.466,29.238,22.176,30.624,32.868,35.574],
      [10.098,2.97,6.93,8.316,11.814,7.458,11.286,7.854,17.094,
      19.272,12.474,21.78,23.76,26.928],
      [16.236,6.93,2.97,8.118,6.996,10.428,12.078,14.85,10.164,
      13.332,6.072,14.784,16.83,19.998],
      [17.952,8.316,8.118,2.97,6.27,4.026,3.96,14.124,13.794,
      13.53,9.108,19.206,20.526,24.354],
      [21.846,11.814,6.996,6.27,2.97,10.23,9.042,19.074,7.722,
      7.655,3.564,13.134,14.322,18.15],
      [15.642,7.458,10.428,4.026,10.23,2.97,4.422,11.022,17.688,
      17.556,12.87,23.034,24.486,28.248],
      [19.998,11.286,12.078,3.96,9.042,4.422,2.97,15.378,16.764,
      15.312,12.408,22.176,23.166,27.126],
      [5.478,7.854,14.85,14.124,19.074,11.022,15.378,2.97,24.948,
      26.664,20.196,29.634,31.68,34.848],
      [26.466,17.094,10.164,13.794,7.722,17.688,16.764,24.948,
      2.97,5.808,4.95,5.412,6.93,10.56],
      [29.238,19.272,13.332,13.53,7.656,17.556,15.312,26.664,5.808,
      2.97,7.326,9.042,8.91,12.87],
      [22.176,12.474,6.072,9.108,3.564,12.87,12.408,20.196,4.95,
      7.326,2.97,10.164,11.748,15.378],
      [30.624,21.78,14.784,19.206,13.134,23.034,22.176,29.634,5.412,
      9.042,10.164,2.97,2.508,15.378],
      [32.868,23.76,16.83,20.526,14.322,24.486,23.166,31.68,6.93,
      8.91,11.748,2.508,2.97,4.026],
      [35.574,26.928,19.998,24.354,18.15,28.248,27.126,34.848,
      10.56,12.87,15.378,5.28,4.026,2.97] ]

  def fire(self) :
    Value = float(self.Value.get(0).toString())
    To = int(self.To.get(0).toString())
    From = int(self.From.get(0).toString())
    calculatedcost = Value*(self.costToFromKg[From][To]+2.97)
    self.Cost.send(0,ptolemy.data.DoubleToken(calculatedcost))
    self.FromOut.send(0,ptolemy.data.IntToken(From))
    self.ToOut.send(0,ptolemy.data.IntToken(To))
    return
```

The code should not be difficult to understand. It essentially has an initialiser (__(self)__), which in this particular case is the loading of the cost table into memory (there may be other unexplored options for doing this within Ptolemy II of which the authors are unaware at the time of writing), and a `fire(self)` method that executes the main code. The `fire` method essentially reads inputs from input ports by calling:

- `self.Value.get(0).toString()`
- `self.To.get(0).toString()`
- `self.From.get(0).toString()`

and then calculates a cost (reading the table for the corresponding costs) and outputs the calculated value using `self.Cost.send (0,ptolemy.data.DoubleTo ken(calculatedcost))`. The `2.97` that is added to the cost for a given origin and destination is to account for the transportation cost within the province to get the products from farmers to distribution centres.

When the model is run, the output that it generates is shown in Fig. 5.18. This graph describes the evolution of the accumulated cost in time. A similar change could be introduced to account for greenhouse emissions in the model; this potential extension is left to the reader.

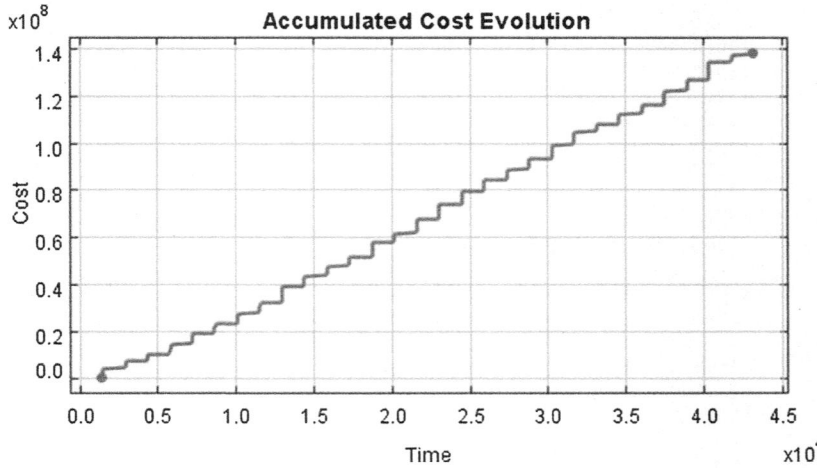

Fig. 5.18 Output of the first case study after being run

5.4 Second Case Study: The Impact of Restructuring the Supply Chain

The second case study looks at the impact of restructuring on the supply chain. The scenario is as follows: "If a factory is added in Ranong province, what changes will occur in terms of transportation costs for the overall network?" For this purpose, the model developed earlier is modified accordingly to incorporate some factories in Ranong and to see what impact it has on the model.

If two factories are added to Ranong, the resulting demand on each province will be distributed as per Table 5.12. As explained, these demands are assigned based on the province weight which is determined by the number of factories each province has.

The first case study model was run 40 times, as was the second case study model. The idea is to determine whether there is any difference between the operation of the supply chain in the two cases. It can be seen that it is better for the supply chain not to add factories in Ranong, as Ranong in the original configuration was mainly providing Chumporn with pri-

Table 5.12 Factory demand data for the second case study (adjusted by author, based on data from (DAT 2016))

Province	LCT (kg)	RSS (kg)	STR (kg)
Chumporn	83,723.735	73,637.646	202,563.306
Suratthani	234,426.457	206,185.408	567,177.256
Nakhon Si Thammarat	343,267.312	301,914.347	830,509.554
Krabi	66,978.988	58,910.116	162,050.645
Trang	200,936.963	176,730.349	486,151.934
Phangnga	25,117.12	22,091.294	60,768.992
Phuket	25,117.12	22,091.294	60,768.992
Ranong	16,744.747	14,727.529	40,512.661
Songkhla	678,162.251	596,464.929	1,640,762.777
Satun	16,744.747	14,727.529	40,512.661
Patthalung	92,096.108	81,001.41	222,819.636
Pattani	75,351.361	66,273.881	182,306.975
Yala	100,468.482	88,365.175	243,075.967
Narathiwat	41,861.867	36,818.823	101,281.653

mary rubber products. If Ranong has its own factories, this will force Chumporn to acquire the primary rubber products from other provinces farther away, which will incur higher transportation costs. This small case study illustrates the type of counter-intuitive insight that DES models allow. If a test is run for differences in means, it can be concluded that the means are statistically significantly different with 95% greater confidence in the costs of case study two than those of case study one. For the sake of completeness, the results of the test (using GRETL[2]) are provided below:

```
Null hypothesis: Difference of means = 0

Sample 1:
 n = 40, mean = 1.30775e+008, s.d. = 4.18922e+006
 standard error of mean = 662374
 95\% confidence interval for mean: 1.29436e+008 to 1.32115e+008

Sample 2:
 n = 40, mean = 1.33405e+008, s.d. = 4.16122e+006
 standard error of mean = 657947
 95\% confidence interval for mean: 1.32075e+008 to 1.34736e+008

Test statistic: t(77) = (1.30775e+008 - 1.33405e+008)/933613 =
-2.81704
Two-tailed p-value = 0.006155
(one-tailed = 0.003077)
```

A graphical representation of the test is provided in Fig. 5.19. It can be seen that the difference between the means is statistically significantly different from zero, and as the results of the test indicate, the p-value is so small that the null hypothesis (difference of means equal to zero) cannot be accepted.

5.5 Summary

This chapter presents the application of the model developed in the previous chapter. It is a complex task due to the multitude of information required and the constant changes/additions to the basic components developed previously. The chapter has illustrated the type of problems that practitioners can run into when developing models. The natural con-

5.6 Note Regarding Ptolemy II Images Used in the Text

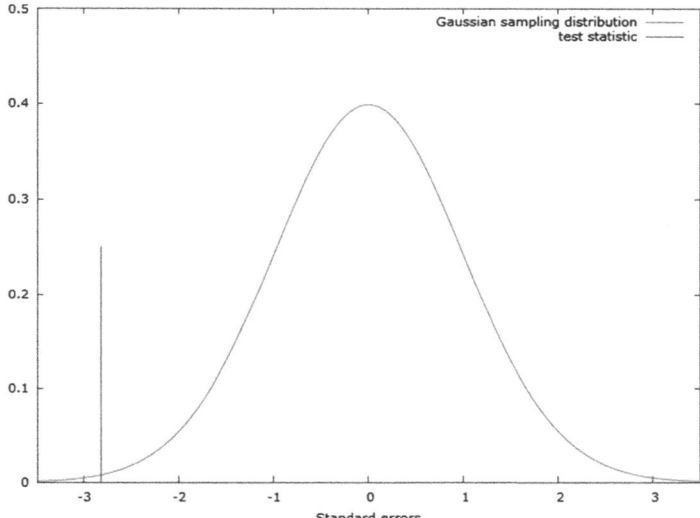

Fig. 5.19 Output of the first case study after being run

clusion at this point is that additional data are required to obtain a better description of the phenomenon. Additional data can, in turn, change the way the model is built and hence the way the basic elements of the model are developed. Once a base case was decided, it was subsequently used to look at case studies that illustrate the type of analyses that are possible with a DES model. It can be observed that the model can be extended to accommodate other analyses not initially contemplated when designing the model.

5.6 Note Regarding Ptolemy II Images Used in the Text

The Ptolemy II images and screen-shots used in the chapter are subject to the following license:

Copyright ©1990–2017 The Regents of the University of California. All rights reserved.

Permission is hereby granted, without written agreement and without license or royalty fees, to use, copy, modify, and distribute this software and its documentation for any purpose, provided that the above copyright notice and the following two paragraphs appear in all copies of this software.

IN NO EVENT SHALL THE UNIVERSITY OF CALIFORNIA BE LIABLE TO ANY PARTY FOR DIRECT, INDIRECT, SPECIAL, INCIDENTAL, OR CONSEQUENTIAL DAMAGES ARISING OUT OF THE USE OF THIS SOFTWARE AND ITS DOCUMENTATION, EVEN IF THE UNIVERSITY OF CALIFORNIA HAS BEEN ADVISED OF THE POSSIBILITY OF SUCH DAMAGE.

THE UNIVERSITY OF CALIFORNIA SPECIFICALLY DISCLAIMS ANY WARRANTIES, INCLUDING, BUT NOT LIMITED TO, THE IMPLIED WARRANTIES OF MERCHANTABILITY AND FITNESS FOR A PARTICULAR PURPOSE. THE SOFTWARE PROVIDED HEREUNDER IS ON AN "AS IS" BASIS, AND THE UNIVERSITY OF CALIFORNIA HAS NO OBLIGATION TO PROVIDE MAINTENANCE, SUPPORT, UPDATES, ENHANCEMENTS, OR MODIFICATIONS.

Notes

1. On the particular machine where this was tested the time went down from around 160 s to 13 s, i.e. a linear decrease.
2. GRETL is open source software for econometric analysis that can be obtained from http://gretl.sourceforge.net/.

References

DAT. (2016). *Thailand rubber statistics*. Thailand: Department of Agriculture.

Shipping, S. (2012). *Costs of rubber transportation*. Hatyai, Thailand: Starlight Express Shipping Company.

TTRA. (2016). Thailand natural rubber production 2010–2016 statistics. The Thai Rubber Association.

6
Conclusions and Future Research Avenues

Abstract This chapter concludes with summaries and recommendations for future research.

Keywords Conclusion · Recommendation · Future research

6.1 Conclusions

Rubber has played and will play an important role in the further development of civilisation. Rubber is important not only because of the prominent role it has in our world but also because of the important social role it plays. Rubber is similar to other renewable natural resources in the nature of extraction and processing. However, the choice of rubber in this book was made to illustrate the type of problems a modeller could face if a discrete event simulation model is sought.

The book has presented the reader with sufficient information to gain a reasonable level of understanding of how the supply chain for the Thai rubber industry currently operates. However, given the complexity of the supply chain and the myriad different rubber products, it was necessary to simplify the presentation. The rationale for doing this was to avoid overwhelming the reader with detail that might drive attention to infrequent phenomena in the supply chain.

The adoption of an open source tool for modelling is expected to attract new practitioners and researchers to the area. As far as the authors know, there have been no previous discrete event simulation models applied to the rubber supply chain. There are some applications of simulation in the agro-industry area, but none of them focus on the particular details pertaining to the rubber supply chain. From this point of view the book presents a unique approach to the supply chain. It is expected that the technique might be expanded from what is presented here, and that further discussion will be conducted for the benefit of those who are involved in the industry.

The book does not aim to provide an encyclopaedic treatment of the topic. On the contrary, it aims to summarise the types of decisions that a practitioner needs to take when attempting to use the DES paradigm to build a model. There are always several decisions that need to be taken at every stage of the process, so it is difficult to provide an infallible "recipe" that can be followed at all times. In the development of the model, what was initially implemented had to be changed at times to accommodate additional information, or just simply to produce a model that could provide the level of interaction required between components. In addition, it must be noted that different tools provide different capabilities, and the choice of Ptolemy II had consequences as to what was feasible in terms of modelling.

One of the outcomes of this investigation is that it has proven that it is possible to use a DES method for modelling Thailand's rubber supply chain. This was achieved through a hands-on approach that allowed the authors to present the ideas in such a way that the reader could participate in building the model, rather than simply being an observer. The model may not be infallible, but therein lie future challenges. Any model is an abstraction and thus can be improved. For the model presented, some obvious improvements are already envisioned, such as access to better data and construction of more realistic models. The next section will discuss some future research paths that could contribute to improving what has been presented in this book.

6.2 Future Research Avenues

The model developed in this book is by no means comprehensive, in the sense that the supply chain to be modelled has numerous small details which create complexity (from a modelling point of view). There are many ways that this particular work might have been carried out. However, the current model is a more than adequate representation of a real supply chain, and allows us to obtain conclusions and analyse "what if" scenarios. The following subsections discuss potential improvements to the current model.

6.2.1 Incorporation of Farm Production Figures

The current model, as implemented, generates the production of a province based on a power law distribution for the size of the farm. This modelling decision was adopted due to lack of sufficient data, along with the impact it could have on the execution speed of the model. Too many tokens being generated could severely impact the execution speed, and it is not clear whether this would have a great impact on the quality of the results obtained by the simulation.

The currently chosen approach has some benefits in that it simplifies the data management process. Everything is summarised in the distributional characteristics of the production, based on farm size and average productivity per unit of land. However, in a more realistic setting, where a particular question is in need of a particular answer, it would be desirable to have a detailed list of all the farmers and, ideally, the types and everyday quantities of their production. This information would certainly allow a more precise approximation of the production volume of a province or region within a province.

6.2.2 Seasonal Influence on Farm Productivity

The productivity of the rubber tree is affected by the seasons. Consequently, seasonal patterns for the production of rubber could be introduced into the model. This would affect the output of each individual farm, which would then translate into an aggregated production for each province.

This factor is not a minor one, particularly when facing external demand, which is characterised as stochastic (as with the current model). A proper characterisation of the seasonal nature of rubber productivity is certainly important in forecasting productivity, which in turn can be used to negotiate future sales and thus reduce the uncertainty associated with pricing and other market factors.

6.2.3 Introduction of a Pricing Model

Rubber is a commodity which is transacted in the rubber trading market. The markets operate under the influence of supply/demand relationships that are difficult to foresee, as they depend on weather and local decisions. Each actor in the supply chain makes decisions that affect the equilibrium price on any given day, first at a local level (distribution centre, cooperative, etc.), and then, in an aggregated way, on the overall price at the market.

Sales by factories of secondary rubber products pull some primary products from the supply chain, making the relative pricing of products difficult to predict. This potential research area is important, as the stability of the supply depends on production and pricing decisions that are usually observed after the fact, but that are important and define the operation of the supply chain.

6.2.4 Implementation of Smart Actors for Implementing Dispatching Decisions

The products that flow in the supply chain are transported. Most farmers do not have many options in deciding where to send their product, so it is usually delivered to the closest distribution centre. This may not be the best possible way of analysing decision options. If a farmer lives in the overlap area of two distribution centres, and one of the centres offers the farmer a better price than the other centre, but the farmer has a longer commercial relationship with the former, a difficult decision must be made. Similarly, to which factories does a distribution centre sell their primary rubber products? Other questions to consider are: Who is in charge of the transportation costs? Do the actors in the system consider relative pricing and delivery costs at all?

At the moment, the distribution centres are modelled as a single entity per province. However, there are several of them, and most likely each one has its own pricing policy, thus potentially affecting factors such as the availability of material for delivery to factories. These decisions, therefore, are far from trivial, and are currently not covered by the DES model.

6.2.5 Implementation of Smart Actors for Implementing Production Decisions

Production decisions are those that answer to external demand. Given an external demand, factories will respond by defining what to do next to meet that demand. This requires decisions on the volume of primary rubber products to buy from distribution centres, how much to have in inventory, and the frequency with which orders are put in to distribution centres. The economic ordering quantity (EOQ) model is the tool typically used to define the optimal frequency, with the objective of minimising the cost of ordering and the holding costs of the operations.

It can be noted that this research path requires the use of optimisation tools currently unavailable in Ptolemy II, and possibly unavailable in any DES package. Simpler strategies could be implemented to see the impact they have on the operation of the system and how much unsatisfied demand is produced by different ordering policies. An interesting associated problem would be to see how the uncertainty regarding external orders might impact upon the supply chain and how this uncertainty propagates.

6.2.6 Implementation of Similar Models in Other Rubber Producing Countries

Thailand is not the only country that produces rubber products. Other producing countries could benefit from the same techniques implemented in this book.

6.2.7 Realistic Simulation of Material Flow in the Supply Chain Model

The model currently implemented produces transactions once a day to avoid overpopulating the future event list. As good as the idea is from a computational point of view, it is a great assumption, as farmer transactions do not occur all at the same time, but throughout the day. The same applies to distribution centres and their relationships to factories. A proper evaluation of the benefit that such a modelling option brings to the model is required to assess if a more realistic simulation of material flow is worth carrying out.

6.2.8 Modelling Different Transportation Modes for Outbound Logistics

The current model makes no provision for outbound logistics. This topic has complexities of its own, and potentially requires more detailed consideration, as it has an impact on the operation of the supply chain, and also impacts on the internal transportation decisions that are taken by the actors.

A
Farmer Size Distributions

This appendix contains the histograms and plots resulting from the Power Law distribution performed for the sample data available for farmer sizes in the different provinces.

A.1 Chumporn Province

See Figs. A.1 and A.2.

140   A   Farmer Size Distributions

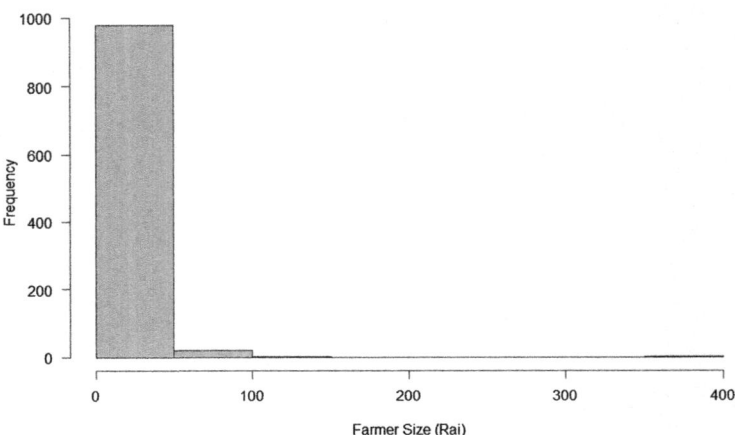

Fig. A.1 Chumporn farmer size histogram

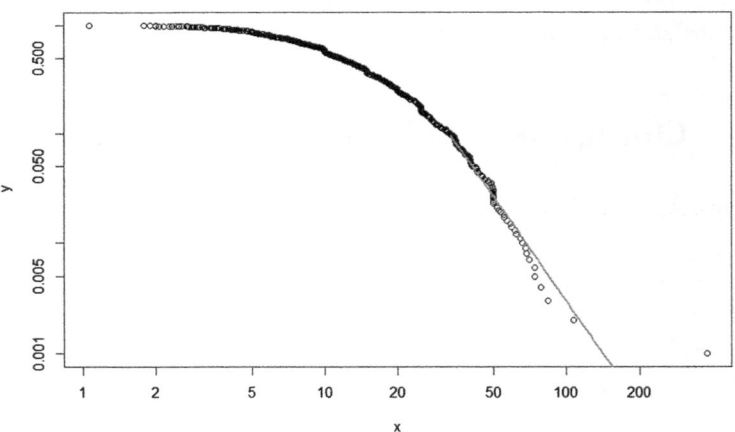

Fig. A.2 Chumporn Province power law fit

A.2 Suratthani Province

See Figs. A.3 and A.4.

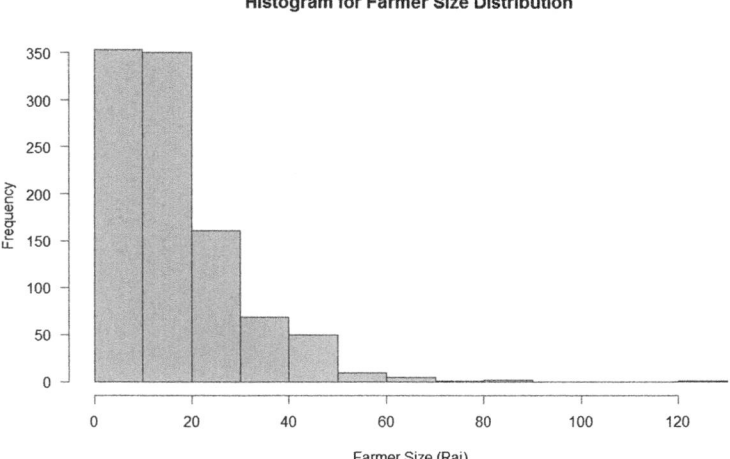

Fig. A.3 Suratthani farmer size histogram

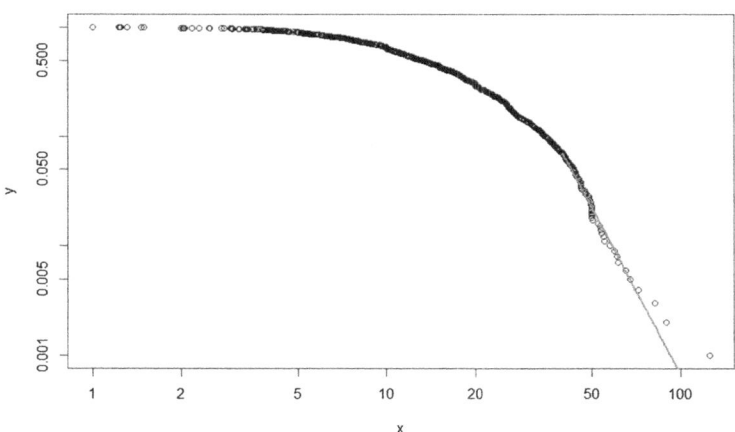

Fig. A.4 Suratthani Province power law fit

A.3 Nakhon Si Thammarat Province

See Figs. A.5 and A.6.

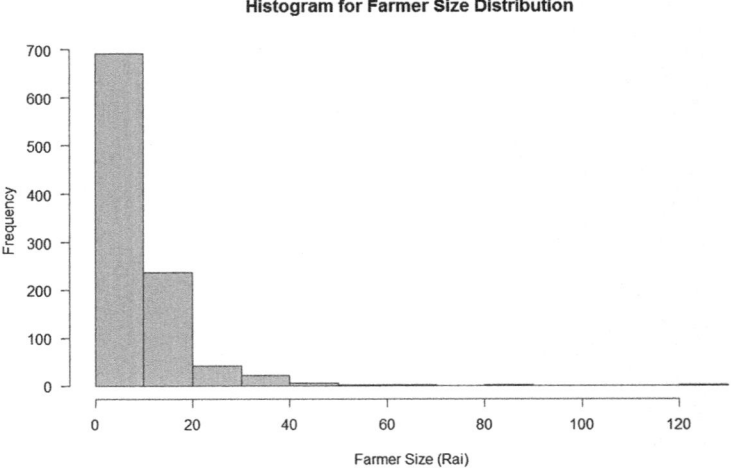

Fig. A.5 Nakhon Si Thammarat farmer size histogram

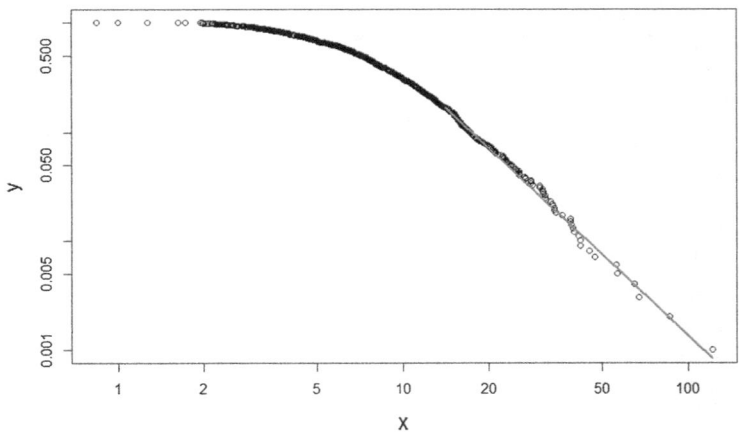

Fig. A.6 Nakhon Si Thammarat Province power law fit

A.4 Krabi Province

See Figs. A.7 and A.8.

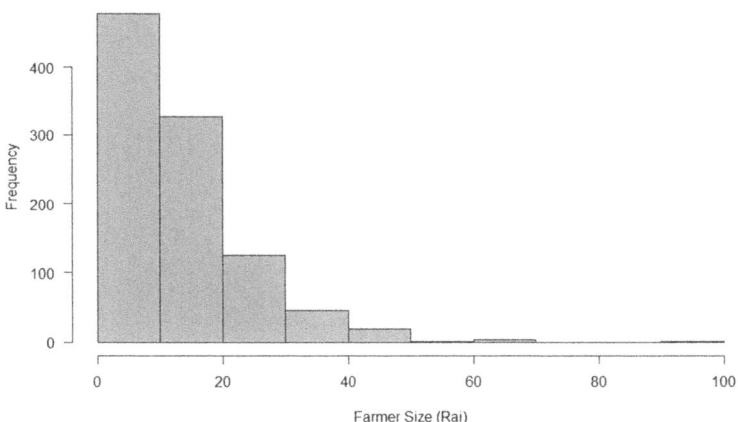

Fig. A.7 Krabi farmer size histogram

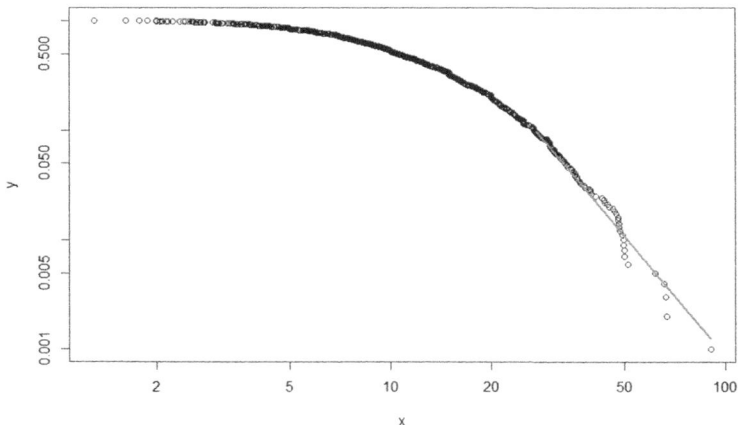

Fig. A.8 Krabi Province power law fit

A.5 Trang Province

See Figs. A.9 and A.10.

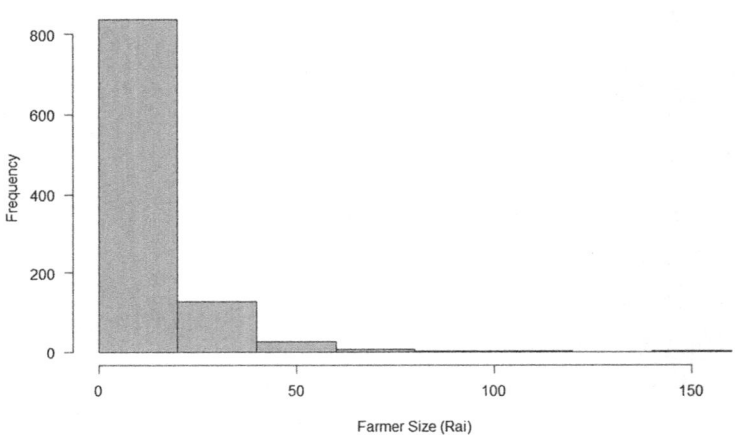

Fig. A.9 Trang farmer size histogram

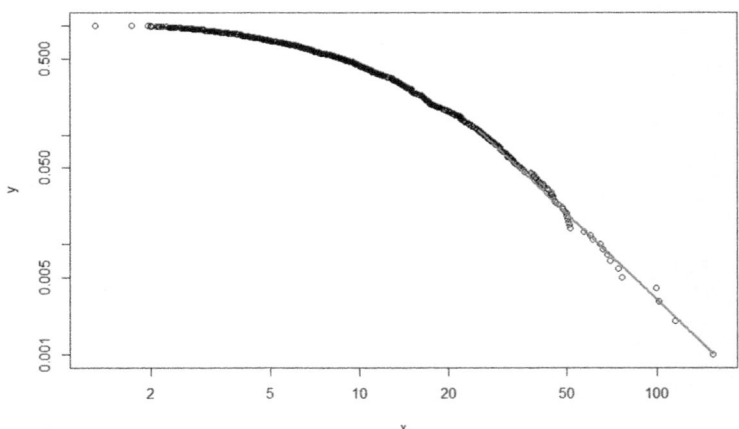

Fig. A.10 Trang Province power law fit

A.6 Phangnga Province

See Figs. A.11 and A.12.

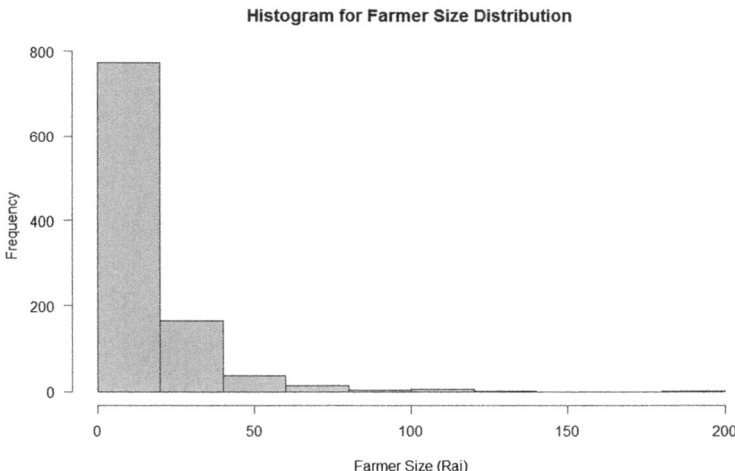

Fig. A.11 Phangnga farmer size histogram

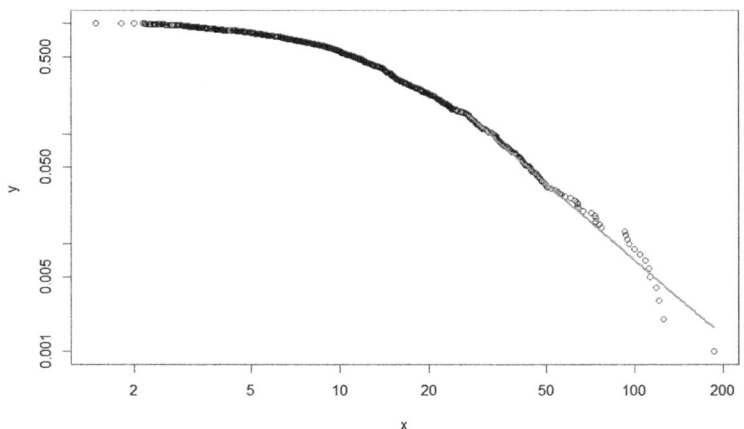

Fig. A.12 Phangnga Province power law fit

A.7 Phuket Province

See Figs. A.13 and A.14.

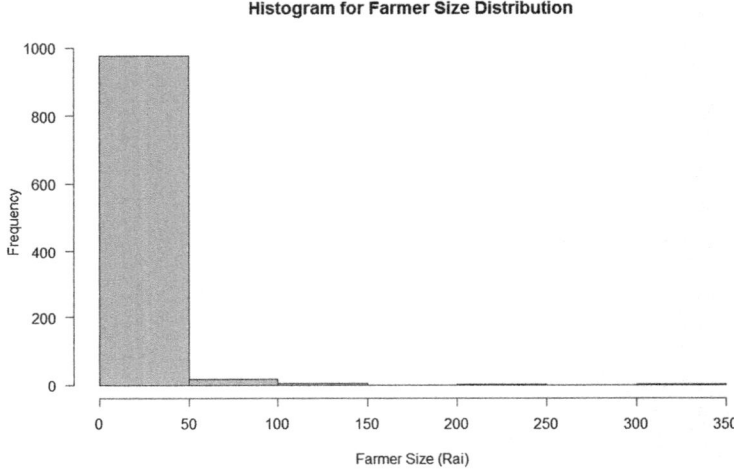

Fig. A.13 Phuket farmer size histogram

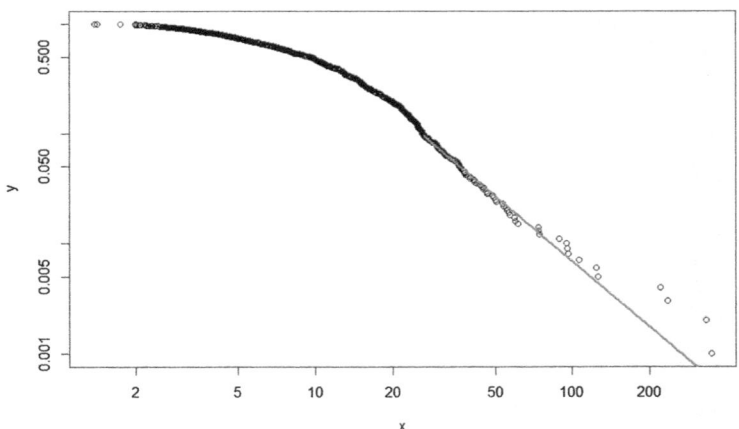

Fig. A.14 Phuket Province power law fit

A.8 Ranong Province

See Figs. A.15 and A.16.

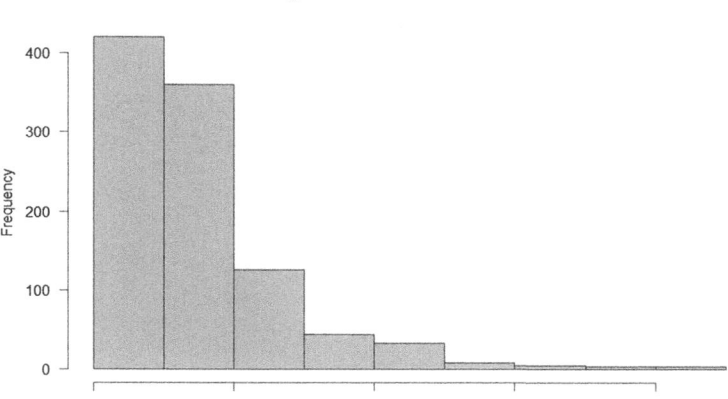

Fig. A.15 Ranong farmer size histogram

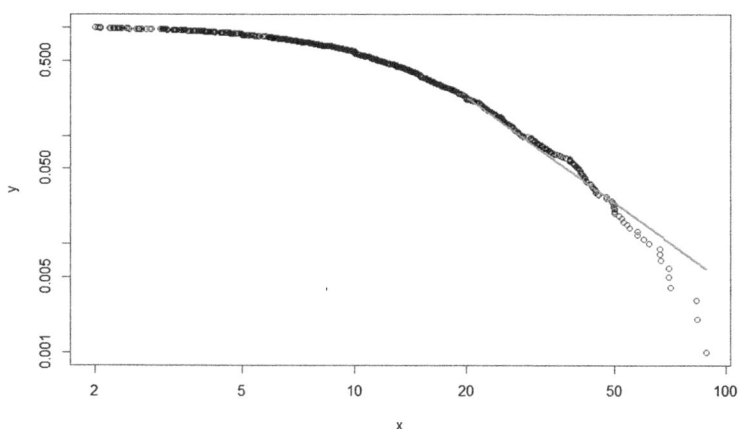

Fig. A.16 Ranong Province power law fit

A.9 Songkhla Province

See Figs. A.17 and A.18.

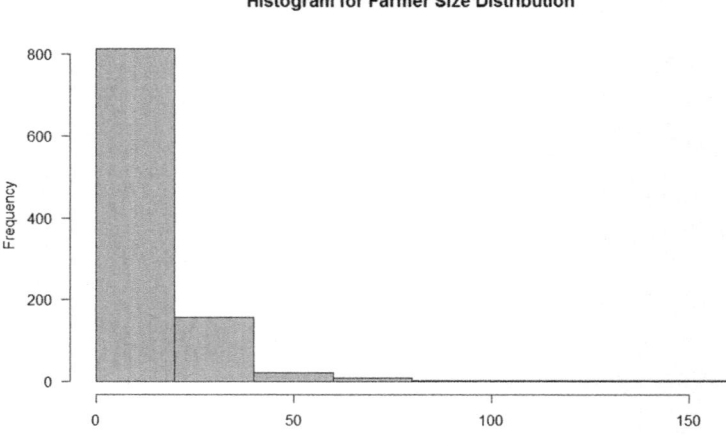

Fig. A.17 Songkhla farmer size histogram

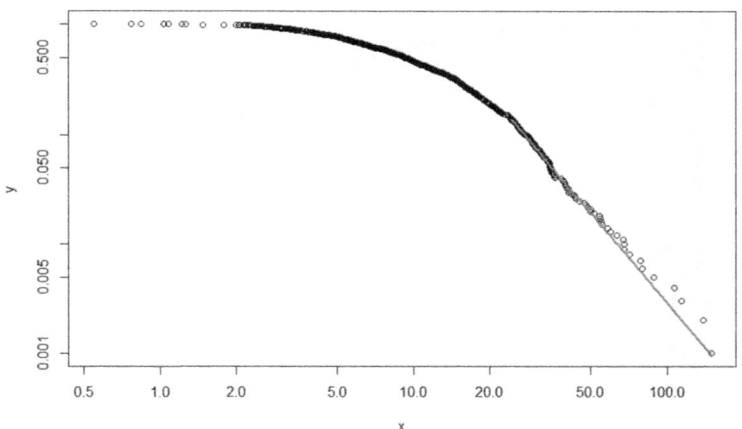

Fig. A.18 Songkhla Province power law fit

A.10 Satun Province

See Figs. A.19 and A.20.

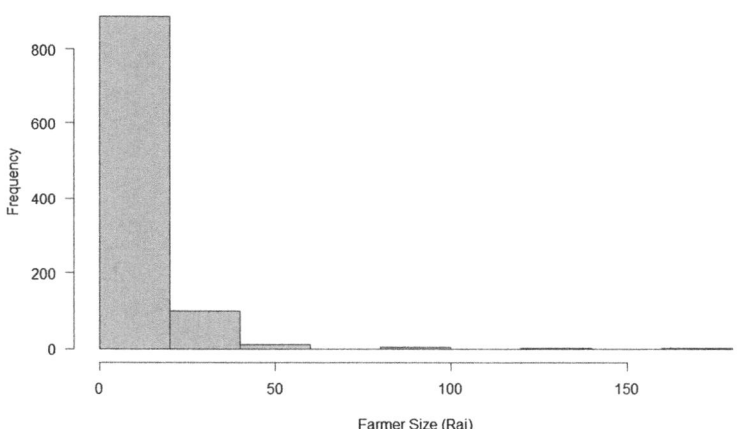

Fig. A.19 Satun farmer size histogram

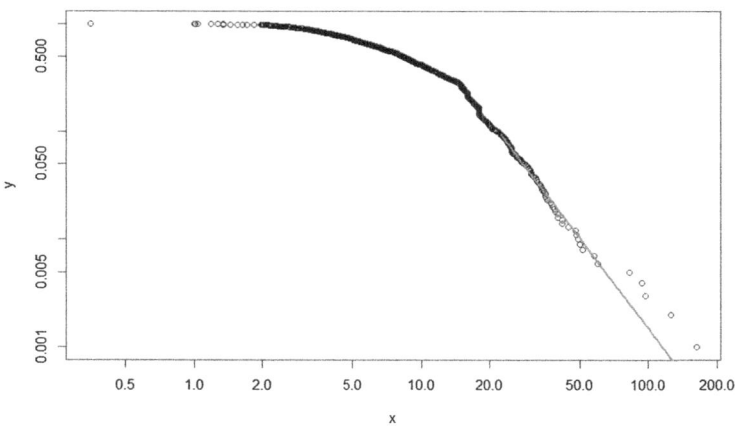

Fig. A.20 Satun Province power law fit

A.11 Pathalung Province

See Figs. A.21 and A.22.

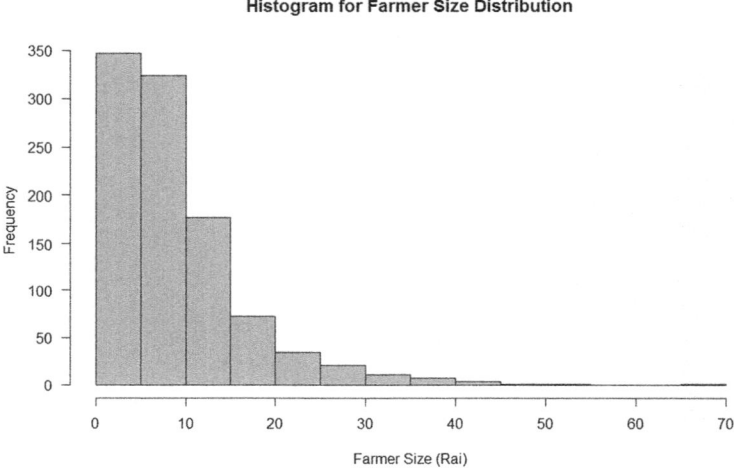

Fig. A.21 Pathalung farmer size histogram

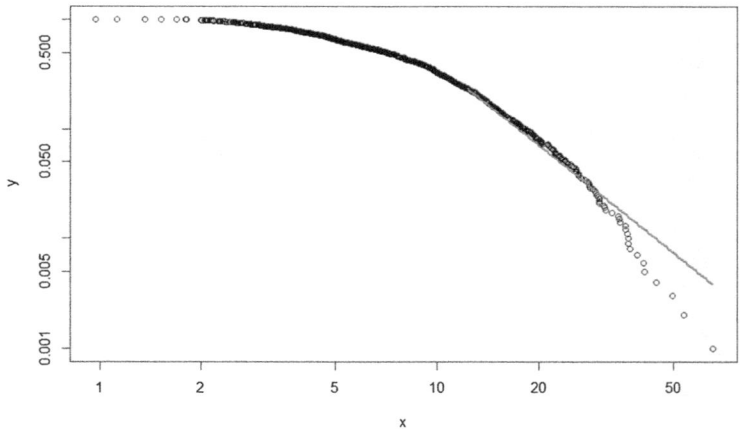

Fig. A.22 Pathalung Province power law fit

A.12 Pattani Province

See Figs. A.23 and A.24.

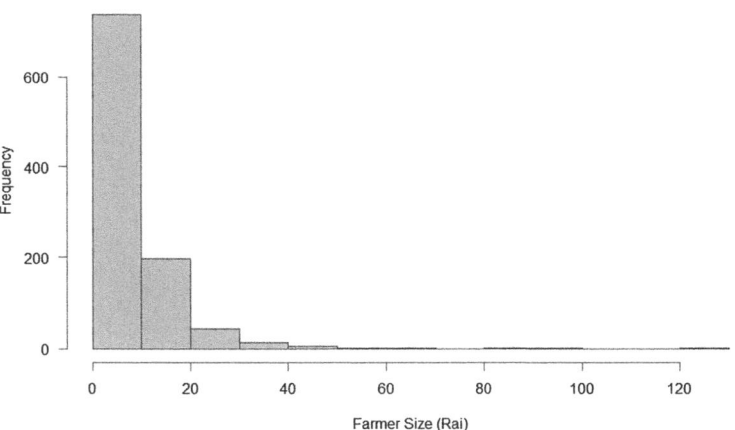

Fig. A.23 Pattani farmer size histogram

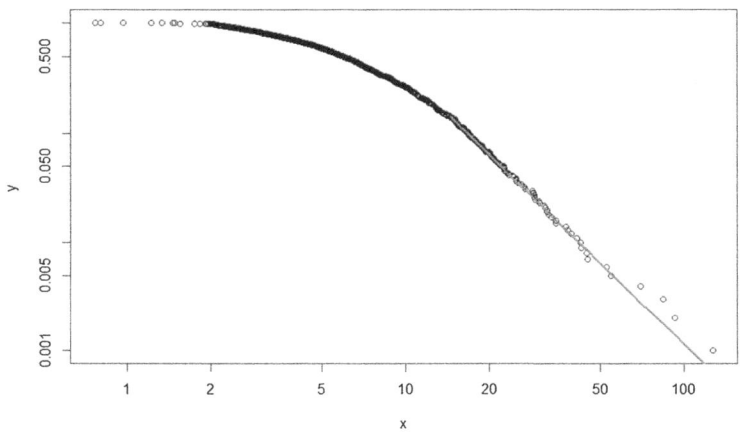

Fig. A.24 Pattani Province power law fit

A.13 Yala Province

See Figs. A.25 and A.26.

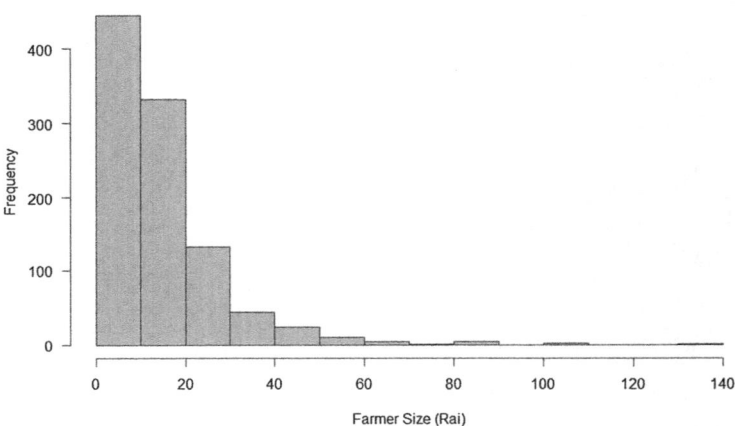

Fig. A.25 Yala farmer size histogram

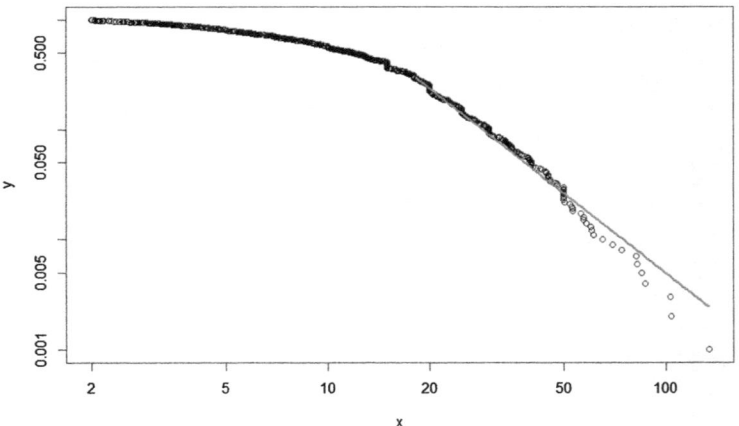

Fig. A.26 Yala Province power law fit

A.14 Narathiwat Province

See Figs. A.27 and A.28.

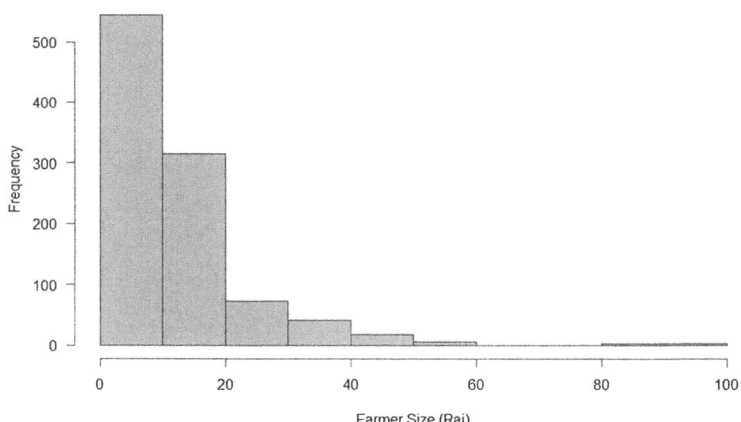

Fig. A.27 Narathiwat farmer size histogram

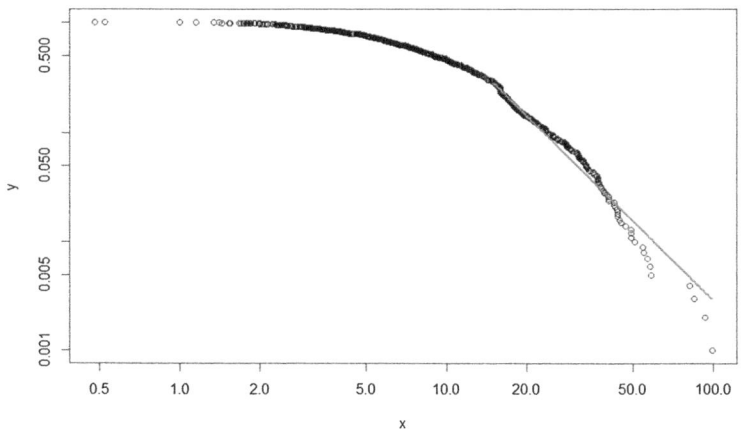

Fig. A.28 Narathiwat Province power law fit

Index

D
Decision Support Systems
　Overview, 9
DES Model, 73
　Changing Distribution Centres, 111
　CL Load, 80
　CL Proportion, 75, 79
　Demand Data (Factories to Provinces), 118
　Demand Modelling, 112
　Dispatcher, 110
　Distribution Center Capacity, 80
　Distribution Centres, 105
　Expressions, 75
　Factories, 107
　Factory Demand Data, 117
　Farm Productivity, 79
　Improving Factory Composite Actor, 108
　LX Load, 80
　LX Proportion, 73, 79
　Number of Farmers, 75
　Parameters, 73
　Productivity per Rai, 75
　Province Load Generator Composite Actor, 81
　Size of Farming, 75
　Stockpile Composite Actor, 81
　　Actual Level, 81
　　Capacity, 81
　　Inflow, 81
　　Initial Level, 81
　　OutFlow, 81
　　reset, 81
　Transportation Costs (between and within Provinces), 125
　US Load, 80
　US Proportion, 75, 79
Discrete Event Simulation, 47, 65
　Mathematical Definition, 48
　World Views, 48
　　Activity Scanning, 49
　　Event Scheduling, 49
　　Process Interaction, 50
Discrete Event System, 44

Index

F
Farmer, 70

H
Heuristics
  Overview, 7

O
Optimisation
  Model for Feasible Assignation, 119
  Overview, 8

P
Poisson Distribution, 55
Power Law Distribution, 70, 71
Ptolemy II, 50, 65
  Graph Editor, 50
  Queue and Server Example, 52
  Tokens, 69
  Tour of Ptolemy II, 52
  Vergil, 50
Ptolemy II Actors
  Accumulator, 81, 103
  AddSubtract, 87
  Array Sum, 76
  Average, 101
  Colt Exponential, 61
    Lambda, 61
  Composite Actor, 70, 72
    Open Actor, 70
  Custom Icon, 83
  DE Director, 54, 67, 101, 103
    Commit, 67
    Stop Time, 55, 67
  Discrete Clock, 68, 69, 88, 103, 116

Discrete Random, 109
Discrete Random Source, 109, 121
Discrete Time Delay, 89
Display, 76, 77
Domain Specific, 59
Expression, 75, 81, 87, 88
  AddSubtract, 81
Gaussian, 75, 79, 87, 89
Generic Sources, 55
Iterate Over Array, 76, 77, 79
Monitor Value, 68, 103
Poisson Clock, 55, 88, 90
  Values, 55
  Mean Time, 55
  Private Seed, 56
  Seed, 55
Python, 124
Queue, 54, 59
Ramp, 57, 68
  Init, 58
  Step, 58
Random Folder, 72
Record, 84
Record Assembler, 90, 111, 124
Record Dissasembler, 85, 87, 124
Repeat, 75, 76
repeat, 76
SDF Director, 77, 78
Sequence Sources, 55
Sequence to Array, 76, 77
Server, 59
Server Actor, 54
Set Variable, 75–77
Switch, 110, 111
Timed Plotter, 56, 60, 76, 91, 113
Timed Sinks, 56

Timed Sources, 55
Uniform, 72, 75, 77
   Lower Bound, 72
   Upper Bound, 72
Utilities, 70

R
Random Number Generator, 46
   Cycle, 47
   Increment, 47
   Linear Congruential Generator, 47
   Modulus, 47
   Multiplier, 47
   Seed, 47
Rubber, 1
   Commercial Tree, 24
   Demand, 3
   Price, 4
   Supply, 4
   Tree Maturity, 26
   Types, 1
   Uses, 1
      Medical Gloves, 3
      Tyre, 3
Rubber Supply Chain
   Air-Dried Sheet, 20
   Cooperatives, 31
   Crepe Rubber, 20
   Cup-Lump, 20
   Distribution Channels, 21
   End Users, 21
   Estate Rubber Farms, 23
   Factories, 21
   Farms, 20
      Income Distribution, 24
   Field Latex, 20
   Final Rubber Production, 35
   First-Tier Market Traders, 20, 30
   Fresh Latex, 20
   General Market, 31
   High-Grade Block Rubber, 20
   Intermediate Rubber Production, 32
      Block Rubber, 33
      Latex Concentrate, 34
      Logistics, 34
      Marketing, 35
      Ripped-Smoked Sheet, 33
   Intermediate Rubber Products, 20
   Latex Concentrate, 20
   Logistics, 27
   Marketing, 28
   Physical Transformation, 20
   Planting, 24
      Density, 24
   Primary Rubber Processing, 26
   Primary Rubber Product Trader, 30
      Logistics, 31
      Marketing, 32
   Ripped-Smoked Sheet, 20
   Second-Tier Traders, 21, 30
   Small Holding Farms, 23
      Sizes, 23
      Proportions, 24
   Standard Block Rubber, 20
   Supply-Demand Mechanisms, 36
      Demand Factors, 38
      Rubber Price Formulation, 38
      Supply Factors, 36
   Tapping, 26
      Extension, 26

Span, 26
Timing, 26
Third-Tier Market Trader, 21, 30
Unsmoked Sheet, 20
Vertical Integration, 30

S
Simulation, 41
    Advantages, 43
    Disadvantages, 43
    Overview, 5
    Study Steps, 45
    Uses, 42
Southern Provinces of Thailand, 98
System
    Activity, 45
    Attribute, 45
    Continuous, 44
    Deterministic, 43
    Discrete, 44
    Dynamic, 43
    Entity, 45
    Environment, 44
    Event, 45
    State, 45
    Static, 43
    Stochastic, 43

T
Thai Rubber
    Number of Farmers, 100
    Overview, 4
    Plantation Areas, 5
    Production, 2
    Production per Province, 98
    Revenue, 2
    Supply Chain, 10

The manufacturer's authorised representative in the EU is Springer Nature Customer Service Centre GmbH, Europaplatz 3, 69115 Heidelberg, Germany. If you have any concerns regarding our products, please contact ProductSafety@springernature.com

Printed and bound by CPI Group (UK) Ltd, Croydon, CR0 4YY

23/03/2026

02076402-0014